RADICAL POLYMERIZATION

RADICAL POLYMERIZATION

J. C. BEVINGTON

Department of Chemistry, University of Birmingham, England

1961

ACADEMIC PRESS
LONDON and NEW YORK

ACADEMIC PRESS INC. (LONDON) LTD.
17 Old Queen Street
London, S.W.1

U.S. Edition published by

ACADEMIC PRESS INC.
111 Fifth Avenue
New York 3, New York

Library of Congress Catalog Card Number: 61–15057

PRINTED IN GREAT BRITAIN BY
SPOTTISWOODE, BALLANTYNE AND CO. LTD.
LONDON AND COLCHESTER

PREFACE

It is hoped that this book will be of interest to those concerned with the reactions of free radicals of the types found in polymerizing systems. It may also be useful to those whose chief interests lie in the study of the chemical and physical properties of high polymers, since the detailed structures of these materials are governed by the precise natures of the reactions by which they are produced. The intention has been to develop a treatment suitable for readers who, already having some knowledge of the chemistry of free radicals and high polymers, wish to delve more deeply. The size of the book has been restricted deliberately; detailed consideration of the formal kinetics of polymerization processes has been omitted. Attention has been directed particularly towards the mechanisms of the steps in the overall process of polymerization, and the competitions between the various alternative reactions.

I acknowledge my great debt to Sir Harry Melville who stimulated my interest in this branch of chemistry and encouraged me so much. I thank my colleagues for their support and help, and the members of my family for their patience and understanding.

J. C. B.

Birmingham,
February, 1961.

CONTENTS

CHAPTER 1

General Introduction

This book is intended to give a general view of the present state of knowledge in the subject of radical polymerization. To achieve this object, it is unnecessary to consider, for example, every monomer which can be polymerized by a radical mechanism; in all sections, the examples selected for discussion are thought to be representative and illustrative of the various types of behaviour which can be expected. Radical polymerizations act as most valuable sources of information concerning the reactions of free radicals in solution. A polymer molecule can be regarded as a permanent record of the chain reaction in which it was produced. The size, shape and composition of the molecule, and the nature of its end-groups, are all controlled by the relative rates and the mechanisms of the elementary steps into which the whole reaction can be divided. Kinetic studies of a polymerization and examination of the resulting polymer can lead to a comprehensive understanding of the mechanisms of the reactions occurring during the polymerization, and of the factors governing their rates.

The general plan of the book is to deal with the elementary reactions in the same order as they occur during the growth of a polymer molecule. Discussion of the production of radicals (Chapter 2) is followed by an account of reactions which do not affect the number of radicals in the system, viz. reactions of primary radicals with monomer (Chapter 3), the growth of polymer chains (Chapter 4) and transfer reactions (Chapter 5). Finally, there is consideration of processes by which growing radicals are removed from the system, either in pairs (Chapter 6), or singly by reaction with other components of the reaction mixture (Chapter 7). It is assumed that the reader is familiar with the basic concepts of chain reactions and radical polymerizations; thus, in Chapter 3, some of the uses of inhibitors and retarders are described, although detailed discussion of such substances is deferred until Chapter 7. In cases such as this, cross-references are given.

References to original papers are by no means complete, but it is hoped that the most significant and recent have been included, and that these may be useful as starting points for literature searches. Hosts of other relevant papers might have been mentioned; some of them can be dismissed in the light of subsequent work, but omission of others can be justified only by the necessity of keeping the text to a reasonable length and of avoiding the appearance of a catalogue.

Although a proper appreciation of the experimental methods is vital, results are presented and discussed in this book with very little consideration of the practical work involved. It is supposed that the reader is either already acquainted with the techniques or is prepared to read other texts for detailed accounts. It would be advisable, however, to call attention to the fact that the commonest method for measuring rates of consumption of monomer, viz. dilatometry, is not an absolute method and depends upon the use of conversion factors. These factors are not always known with certainty; for a particular monomer, effects of temperature are generally recognized, but the values of the factors may depend also upon the concentration of monomer and the nature of any diluent which may be present.

Methods for evaluating the velocity constants for the individual steps in the overall process of polymerization are not described; this subject has been dealt with adequately by Burnett (1954) and Bamford, Barb, Jenkins and Onyon (1958) and, at present, it is not possible to improve on their treatments. The precise values of the velocity constants for the individual steps in a radical polymerization depend upon the method for defining the velocity constant for the interaction of pairs of similar radicals. Some authors write the rate of this reaction as k_t (concentration of radicals), others as $2k_t$ (concentration of radicals). There is a strong case for adopting the latter convention, but the former is used here, mainly because it is the one used by Burnett and Bamford *et al.*, and frequent references are made to their books in connection with kinetic aspects.

It is disappointing that in spite of the care which has been devoted to kinetic work, there are still some substantial discrepancies between the results of different groups of workers. Values for velocity constants are quoted at various points in this book, but no attempt has been made to achieve complete consistency in the choice. In most cases, the interest lies in the ratio of the velocity constants for competing reactions; these ratios can often be determined by direct experiments, and the various published values are in much closer agreement than those of the individual velocity constants, which can be found only by rather elaborate methods. In other cases, the significant feature is the dependence of velocity constant upon the precise structures of the reactants or upon the environment; under these circumstances the absolute values of the velocity constants are not of paramount importance and it is sufficient to have reliable relative values, preferably determined by a single set of workers. Frequently data on the rates of related reactions are analysed by considering the dependence of log (velocity constant) upon some parameter; if the velocity constants cover a very wide range of values, as is often the case, uncertainty over their exact values hardly affects the general picture.

In some parts of the book, the reader may be given the impression that more refined examinations of radical polymerizations and the resulting polymers merely reveal more intricacies in the process of polymerization. This is true in the sense that it has become possible to recognize the existence of reactions which occur infrequently and were previously undetected; it must be noted, however, that all the reactions can be placed into a small number of classes. It will be shown that it is unusual for a particular step to proceed entirely by a single mechanism although one is much preferred in most cases. The importances of some of the alternative reaction paths must not be exaggerated, but nevertheless they may be very significant because the resulting abnormal groupings in the polymer molecules may act as centres for chemical reactivity.

In constructing a reaction scheme for a kinetic treatment of a radical polymerization, it is not necessary to specify the detailed courses of the various steps, and it appears that sometimes the fundamental chemistry is almost forgotten in the subsequent mathematical treatment. An example of a case where the kinetic approach may be misleading is the interaction of unlike polymer radicals, i.e. one of the termination processes in co-polymerizations and in reactions where transfer is accompanied by pronounced retardation. In a number of systems, it seems that the velocity constant for this inter-radical reaction depends markedly upon the composition of the reaction mixture, and yet it is a fundamental principle of reaction kinetics that this should not be so. It can only be inferred that the reaction scheme, upon which the kinetic analysis is based, is inadequate and that certain significant processes have been omitted.

Finally, it is necessary to call attention to abbreviations and symbols used in this book. Certain chemical and mathematical equations are numbered, starting afresh at the beginning of each chapter; these equations are not necessarily the most important and the numbering is only for reference purposes. Subscripts to velocity constants (k), pre-exponential factors (A) and activation energies (E) refer to these equations; subscripts p and t correspond to the propagation and termination reactions respectively, and f to the first stage in transfer to monomer. The symbol [ ] refers to the concentration of the species in the brackets and, where appropriate, it is the molar concentration. A polymer radical is represented as P·, the subscript n indicating that it has n monomer units; in discussion of co-polymerization, the symbols $P_a\cdot$ and $P_b\cdot$ refer to polymer radicals having terminal monomer units of types a and b

respectively. Small radicals, of the types derived from initiators, are represented as R·. Monomer is indicated by M, but this symbol may refer also to a monomer unit in a polymer chain; subscripts may be added when co-polymerization is being considered. All other symbols and abbreviations are either those commonly used in publications dealing with high polymers, or as defined in the text.

CHAPTER 2

Production of Radicals

A. INTRODUCTION

This chapter is concerned with methods for producing radicals of known structures by thermal and photochemical reactions. Radicals generated by means of high-energy radiations or by mechanical means are not considered; all the valency bonds in a system are susceptible to rupture by these forms of energy and, except in a few cases, a wide variety of radicals is produced.

To give radicals by thermal dissociation at relatively low temperatures, a molecule must contain a weak valency bond such as that between oxygen atoms in peroxides. It is important to know the rate of production of radicals from an initiator; this must be related to the rate of decomposition of the initiator, but some compounds can decompose by more than one mechanism, not all of them producing radicals. In other cases, the rate of decomposition varies considerably from one solvent to another, and in a particular solvent the velocity constant may appear to depend upon the concentration; these are sure signs of induced decomposition in which radicals attack molecules of undissociated initiator.

Even if induced decomposition is absent, there may still be differences between the rates of dissociation of a particular initiator in various solvents; these can be attributed to differences between the extents to which the initial and transition states are solvated. Consider Fig. 2.1 and suppose that I_0 and T_0 are the energy levels for the initial and transition states for the dissociation of an isolated molecule; in solution, both states are solvated so that energies become $(I_0 - \Delta E_i)$ and $(T_0 - \Delta E_t)$. The energy of activation (E_s) for dissociation in solution is therefore related to that for dissociation of an isolated molecule (E_0), thus

$$E_s = E_0 + (\Delta E_i - \Delta E_t)$$

It is quite likely that $(\Delta E_i - \Delta E_t)$ should vary from one solvent to another so that the energies of activation for dissociation also may depend upon the nature of the solvent.

These solvation effects can also account for a relationship between the energies of activation and the A factors for the dissociations of a particular compound in a series of solvents. It is commonly found that a reduction in E is accompanied by a reduction in A, i.e. a reduction in the entropy of activation. A large and negative value for $(\Delta E_i - \Delta E_t)$, because of increased solvation in the transition state, gives a relatively low value for E_s; in passing from the initial to the transition state, however, there must be increasing order in the arrangement of solvent molecules, and this makes for a low value for the entropy of activation. A similar relationship exists between E and A for the dissociations of members of various series of closely related initiators in a particular solvent. A structural modification which reduces E is likely also to reduce A, because the configurational requirements for the transition state become more stringent.

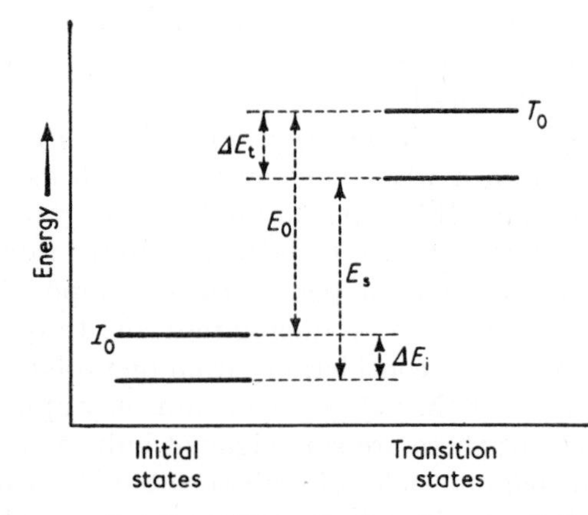

FIG. 2.1. Diagram showing effect of solvent upon E for dissociation of an initiator.

Many thermal sensitizers require an activation energy of roughly 30 kcal/mole for dissociation. The rate of production of radicals varies quite rapidly with temperature, and for each initiator there is a range of about 20°C in which it can be used conveniently. Very low activation energies are usually associated with dissociations giving radicals highly stabilized by resonance; such radicals react only sluggishly with monomers so that initiation is inefficient.

For all the thermal sensitizers which have been examined at high pressures, the rate of the primary dissociation is reduced by increasing the applied pressure. The effect arises because there must be an extension of at least one valency bond in passing to the transition state, giving a positive value for the "volume of activation".

For several series of initiators, the members of which differ only in the substituent at the *para* or *meta* position in a benzene ring, the rate of dissociation is related to the Hammett substituent constant. This constant expresses the capacity of the substituent to donate or accept electrons, and it is deduced that polar factors are significant in the dissociations of these initiators.

In a number of cases, the radicals first formed may subsequently dissociate into smaller radicals and stable molecules. The number of types of radical in the system is thereby increased, but the effect may be of considerable value in two distinct connections, viz.:

(*a*) comparison of the reactivities of monomers and other scavengers towards reference radicals (see Chapter 3, D.*1*);
(*b*) examination of the cage-effect (see Chapter 3, B).

Photo-dissociation of stable molecules is another important method for generating radicals. Many substances which can be used as thermal sensitizers can also function as photosensitizers. Usually the same radicals are produced in the two types of dissociation, but those formed photochemically may not be in their ground states, and their reactivities may therefore be somewhat different from those of radicals generated thermally at the same temperature.

Radicals of defined structure can also be produced by transfer reactions of various types. These may be electron-transfer processes, either thermal or photochemical, in which case the point of particular interest is that the processes require comparatively small energies of activation and can be used to generate radicals at quite low temperatures. Radical-displacements, which include hydrogen or halogen abstraction, can be used to generate particular radicals which are not easily produced by dissociative processes. An important distinction between dissociative and transfer processes for producing radicals, is that the former give pairs of radicals whereas the latter give single radicals.

B. Peroxides

Peroxides form the most important group of sensitizers for radical polymerizations. In the following discussion, the simplest peroxide, viz. hydrogen peroxide, is considered first; then, an account is given of an inorganic peroxide, viz. the persulphate ion. This is followed by discussion of certain of the more important organic peroxides, derived by the replacement of hydrogen atoms in hydrogen peroxide by organic radicals; for more comprehensive accounts, the book by Tobolsky and Mesrobian (1954) and the article by Hawkins (1950) should be consulted.

Hydrogen peroxide has had very limited use as a thermal sensitizer (Nandi and Palit, 1955), but it is best known as a component of redox systems in which radicals are produced by a one-electron transfer process. One such reaction

$$H_2O_2 + Fe^{++} \longrightarrow \cdot OH + OH^- + Fe^{+++} \quad (1)$$

has been studied in great detail by Barb, Baxendale, George and Hargrave (1951) who have also examined the related system in which the ferric ion catalyses the decomposition of the peroxide.

If all the hydroxyl radicals formed in (1) are captured by monomer

$$\cdot OH + CH_2{:}CHX \longrightarrow HO.CH_2.CHX\cdot \quad (2)$$

one ferrous ion is oxidized for each molecule of peroxide consumed. Radicals may also engage in the reactions

$$Fe^{++} + \cdot OH \longrightarrow Fe^{+++} + OH^- \quad (3)$$

$$H_2O_2 + \cdot OH \longrightarrow H_2O + HO_2\cdot$$

If monomer is absent and [peroxide] is much less than [ferrous ion], all the ·OH radicals react according to (3) and one molecule of hydrogen peroxide oxidizes two ferrous ions. In the presence of monomer, (2) and (3) compete and one molecule of peroxide is equivalent to between one and two ferrous ions depending upon the reactivity of the monomer, [monomer], [ferrous ion] and the temperature. This dependence forms the basis for comparison of the reactivities of monomers towards the ·OH radical using (3) as the reference reaction (see Chapter 3, D.*2*).

Persulphates

Various inorganic peroxy compounds have been used as water-soluble initiators in emulsion polymerizations, but only the persulphates have been studied in detail. Thermal dissociation to sulphate radical-ions

$$S_2O_8^{--} \longrightarrow 2SO_4\cdot^- \quad (4)$$

may be accompanied, in acid media, by a catalysed decomposition (Kolthoff and Miller, 1951). The rate of (4) has been measured from the rate of consumption of persulphate (Kolthoff and Miller, 1951) and also by using diphenylpicrylhydrazyl as a radical scavenger to count the radicals being produced (Bawn and Margerison, 1955); the results are not in close agreement, but E_4 is clearly near 30 kcal/mole.

The sulphate radical-ion may be captured by monomer to initiate polymerization. Several groups, including Kolthoff, O'Connor and Hansen (1955), have used ^{35}S-persulphate and claimed that the resulting polymers contain labelled fragments; the radical-ion may, however, react with water

$$SO_4\cdot^- + H_2O \longrightarrow HSO_4^- + \cdot OH \quad (5)$$

and the resulting hydroxyl radical may be the actual initiating species (see Chapter 3, D.*2*).

Persulphates have also been examined as initiators in non-aqueous systems (Sengupta and Palit, 1951; Jarkovsky, Stannett and Szwarc, 1955) using mixtures containing glycols as solvents. The molecular weights of the resulting polymers were higher than expected and it appears that the rates of termination in the reactions were low (see Chapter 6, C).

The persulphate ion is a component of many redox systems, e.g.

$$Fe^{++} + S_2O_8^{--} \longrightarrow Fe^{+++} + SO_4^{--} + SO_4^{\cdot -}$$

for which the energy of activation is only about 12 kcal/mole (Orr and Williams, 1955). Depending on the balance between

$$SO_4^{\cdot -} + M \longrightarrow {}^{-}SO_4.M\cdot$$

and

$$SO_4^{\cdot -} + Fe^{++} \longrightarrow SO_4^{--} + Fe^{+++}$$

one persulphate ion is equivalent to between one and two ferrous ions, and it is possible to compare the reactivities of monomers towards the sulphate radical-ion (see Chapter 3, D.*2*). In the treatment, the possibility of (5) is ignored.

Hydroperoxides

For consideration of the hydroperoxides, the *tert.*butyl and cumene compounds can be taken as typical. The primary dissociation

$$R.O.O.H \longrightarrow R.O\cdot + \cdot OH \qquad (6)$$

may be followed by dissociation of the alkoxy radical (see p. 11). Radical-induced decompositions are quite common and transfer to hydroperoxide may be quite pronounced during polymerizations. The induced decomposition can give peroxy radicals (Martin and Norrish, 1953)

$$(CH_3)_3C.O\cdot + (CH_3)_3C.O.O.H \longrightarrow (CH_3)_3C.OH + (CH_3)_3C.O.O\cdot$$

and these may be formed also during transfer reactions. There is evidence that initiation by hydroperoxides may not be a simple process in which (6) is followed by capture of the radicals by monomer, and that it may involve an interaction of the peroxide with monomer (Walling and Chang, 1954; Schröder, 1958). Hydroperoxides may also decompose by polar mechanisms.

Hydroperoxides are constituents of some of the most important redox initiating systems, e.g.

$$R.O.O.H + Fe^{++} \longrightarrow Fe^{+++} + OH^{-} + RO\cdot \quad \text{or} \quad FeOH^{++} + RO\cdot \qquad (7)$$

Product analysis has shown that alkoxy and not hydroxyl radicals are produced in these reactions (Kharasch, Fono and Nudenberg, 1950;

Kharasch, Arimoto and Nudenberg, 1951). E_7 is usually in the range 9–15 kcal/mole, whereas E_6 for *tert.*butyl hydroperoxide, for example, is about 39 kcal/mole (Bell, Raley, Rust, Seubold and Vaughan, 1951). Information on the reactivities of alkoxy radicals towards monomers can be obtained from study of the competition between

$$R.O\cdot + M \longrightarrow R.O.M\cdot$$

and

$$R.O\cdot + Fe^{++} \longrightarrow R.O^{-} + Fe^{+++}$$

just as in the case of the hydroxyl radical and the sulphate radical-ion (Orr and Williams, 1955).

The thermal and redox dissociations of the dihydroperoxide of meta-di(*iso*propyl)benzene have been used in the preparation of block co-polymers of styrene and methyl methacrylate (Molyneux, 1960). The polymerization of the first monomer is initiated by thermal dissociation

TABLE 2.1

DATA ON REACTIONS BETWEEN PEROXIDES AND THE FERROUS ION

Peroxide	A_7 (mole^{-1} l.$^{+1}$ sec^{-1})	E_7 (kcal/mole)
p.nitrocumene hydroperoxide	$8 \cdot 0 \times 10^{10}$	13·1
cumene hydroperoxide	$1 \cdot 1 \times 10^{10}$	12·0
p.*iso*propylcumene hydroperoxide	$4 \cdot 0 \times 10^{9}$	10·8
p.*tert.*butylcumene hydroperoxide	$1 \cdot 8 \times 10^{9}$	9·9
hydrogen peroxide	$4 \cdot 45 \times 10^{8}$	9·4†

† Results of Barb, Baxendale, George and Hargrave (1951).

of one of the peroxy linkages of the dihydroperoxide so that some of the resulting polymer molecules have hydroperoxide end-groups. These end-groups are subsequently used with ferrous ion to initiate polymerization of the second monomer; the final product includes some polymer molecules consisting of polymethyl methacrylate chains linked to polystyrene chains through $\cdot O.CH_2.C_6H_4.CH_2.O\cdot$ groups.

The rate of the reaction between cumene hydroperoxide and the ferrous ion is affected by substitution in the benzene ring (see Table 2.1) (Orr and Williams, 1955).

For the series of para-substituted cumene hydroperoxides, E_7 can be related to the Hammett substituent constant, σ. It is believed that the first stage in the reaction is the formation of the unstable complex

$$\underset{\dot{H}}{(R.O.O)}\,Fe^{++}$$

and that an electron excess at the oxygen–oxygen bond, caused by presence of an electron-releasing substituent in R, facilitates the subsequent dissociation of the complex to give OH^-. The correlation between E_7 and A_7 for this series of reactions is to be noted.

Peroxy radicals can be formed from *tert*.butyl hydroperoxide thus:

$$(CH_3)_3C.O.O.H + Cu^{++} \longrightarrow (CH_3)_3C.O.O\cdot + H^+ + Cu^+$$

this reaction being followed by

$$(CH_3)_3C.O.O.H + Cu^{+} \longrightarrow (CH_3)_3C.O\cdot + OH^- + Cu^{++}$$

The peroxy radicals can be captured by monomers to give polymers with unstable end-groups (Smets, Poot, Mullier and Bex, 1959).

Di-alkyl Peroxides

Di-alkyl peroxides are more satisfactory than the hydroperoxides as sources of radicals, being less susceptible to induced decomposition and transfer reactions during polymerization. They dissociate to alkoxy radicals

$$R.O.O.R \longrightarrow 2R.O\cdot \qquad (8)$$

For dissociation in solution, values of k_8 for typical di-alkyl peroxides are given by

$$2\cdot 8 . 10^{14} \exp(-35\cdot 0 \text{ kcal}/RT) \text{ sec}^{-1}$$

for di-*tert*.butyl peroxide (Offenbach and Tobolsky, 1957)

and $$4\cdot 31 . 10^{14} \exp(-34\cdot 5 \text{ kcal}/RT) \text{ sec}^{-1}$$

for dicumyl peroxide (Bailey and Godin, 1956).

These peroxides are particularly useful as thermal sensitizers for polymerizations at temperatures between 100° and 120°C. Di-*tert*.butyl peroxide is quite a volatile liquid and has been used as a source of radicals in the gas phase.

The *tert*.butoxy radical may dissociate further:

$$(CH_3)_3C.O\cdot \longrightarrow (CH_3)_2CO + CH_3\cdot \qquad (9)$$

E_9 is about 15 kcal/mole. Dissociation to a ketone and a hydrocarbon radical is a general phenomenon for alkoxy radicals (Gray and Williams, 1959). When more than one type of dissociation is possible, the one giving the largest alkyl radical predominates; thus the butyl radical is formed in preference to the propyl radical, and so on. The phenyl radical is detached from an alkoxy radical with difficulty so that the favoured decomposition of the cumyloxy radical is

$$C_6H_5.C(CH_3)_2.O\cdot \longrightarrow C_6H_5.CO.CH_3 + CH_3\cdot$$

The yield of ketone is depressed if decomposition of a di-alkyl peroxide is performed in the presence of substances which react with alkoxy radicals. This forms the basis of a competitive method for comparing the reactivities of scavengers towards alkoxy radicals (see Chapter 3, D.*1*). Dissociation of the substituted alkoxy radical, $(CH_3)_2C(CN).O\cdot$, is mentioned in Section C of this chapter.

The rate of dissociation of di-*tert.*butyl peroxide is, as expected, reduced by increasing the pressure (Laird, 1956). Dissociation of an alkoxy radical also should be less likely at high pressures, especially if it is competing with a bimolecular process such as addition of the radical to a molecule of monomer.

Russell (1959) has presented evidence for the existence of π-complexes between aromatic hydrocarbons and the *tert.*butoxy radical; in its complexed forms, the radical may exhibit reduced but more selective reactivity. The *tert.*butoxy radicals produced from the peroxide by photo-dissociation in the near u.v. have a reactivity slightly different from that of radicals generated thermally (Allen and Bevington, 1960*a*). If shorter wavelengths are used for the photolysis, the dissociation may take a different course, and at 3000 Å about 10% of the peroxide dissociates thus (Frey, 1959):

$$(CH_3)_3C.O.O.C(CH_3)_3 \longrightarrow 2(CH_3)_3C\cdot + O_2$$

The *tert.*butoxy radical quite readily abstracts hydrogen atoms from stable molecules, and this process can compete quite successfully with addition to olefinic bonds. Thus, at 130°C, the velocity constants for abstraction of hydrogen from *cyclo*hexane and for addition to styrene are comparable (Allen and Bevington, 1960*a*); also, the radical promotes cross-linking of polyisoprenes by dehydrogenation of the polymer, whereas the benzoyloxy radical does so mainly by addition to the double bond (Farmer and Moore, 1951).

Di-acyl Peroxides

The di-aroyl peroxides dissociate thermally to aroyloxy radicals which may themselves subsequently dissociate:

$$R.CO.O.O.CO.R \longrightarrow 2R.CO.O\cdot \longrightarrow 2R\cdot + 2CO_2$$

If reactive substances are absent, there are almost quantitative yields of carbon dioxide from dibenzoyl peroxide (Barson and Bevington, 1958), but scavengers can completely suppress the second stage by reacting with the benzoyloxy radicals (see Fig. 2.2); this confirms that carbon dioxide is not produced in the primary thermal dissociation of the peroxide. During photo-dissociation in the near u.v., however, some carbon dioxide is formed in the primary dissociation (Bevington and

Lewis, 1958), and high concentrations of scavengers fail to suppress completely the formation of carbon dioxide; about 30% of the benzoyloxy radicals which might be formed actually appear as ($C_6H_5\cdot + CO_2$).

For the thermal dissociation of dibenzoyl peroxide at very low concentration in inert solvents, most of the quoted activation energies lie between 29 and 31 kcal/mole, and the A factors between 3×10^{13} and 6×10^{14} sec^{-1}. Some variation of E and A with solvent can be expected because of differences between the capacities of the various solvents to solvate the initial and transition states. This peroxide is best used as an initiator of radical polymerizations between 60° and 80°C.

The behaviour of other di-aroyl peroxides is similar to that of dibenzoyl peroxide. The rate of thermal dissociation is profoundly affected by the

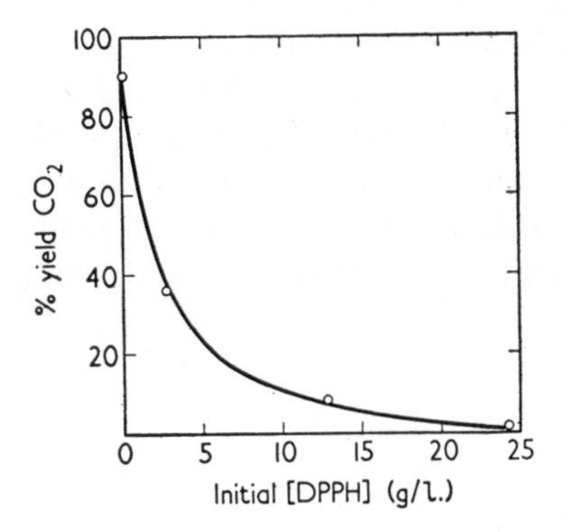

FIG. 2.2. Effect of diphenylpicrylhydrazyl upon yield of CO_2 from dibenzoyl peroxide at 60° C (0·5 g peroxide/l. of benzene; time = 96 hours).

presence of substituents in the benzene rings. Two sets of workers (Swain, Stockmayer and Clarke, 1950; Blomquist and Buselli, 1951) have shown that the data give a good Hammett plot; the values of the velocity constants used in the original plots may need some revision but the general pattern is unaltered. Electron-releasing groups, e.g. CH_3O, accentuate a dipole interaction across the oxygen–oxygen bond in the molecule of the peroxide and so facilitate the dissociation; the opposite effect is exerted by electron-attracting groups such as CN. Substituents in the ring also affect the stability and reactivity of the aroyloxy radicals (Bevington, Toole and Trossarelli, 1958 *a*, *b*); a substituent which causes the peroxide to dissociate rapidly, makes the aroyloxy radical less reactive towards a monomer such as styrene, and also less liable to decompose to the corresponding aryl radical.

Many of the di-aroyl peroxides are susceptible to induced decomposition and to transfer reactions; this is particularly so when the molecule contains electron-attracting groups. Detailed kinetic analyses, notably by Nozaki and Bartlett (1946), have distinguished between the contributions of the direct and induced reactions to the total rate. Specific effects may be very important in these reactions; for example, the polystyrene radical does not attack dibenzoyl peroxide to an appreciable extent but is very reactive towards brominated derivatives of the peroxide, and yet the polymethyl methacrylate radical has little effect even upon the latter type of peroxide (see Chapter 5, E). If the use of several monomers of different types leads to almost the same value for the rate of removal of an initiator, it is reasonable to conclude that the induced decomposition (or transfer to initiator) can be neglected, since it is most improbable that the various polymer radicals are equally effective in attacking undissociated peroxide.

The rate of the direct thermal dissociation of dibenzoyl peroxide is decreased at very high pressures, but the bimolecular induced decomposition is accelerated (Nicholson and Norrish, 1956). Increase of pressure should also tend to suppress dissociation of an aroyloxy radical and so affect the detailed mechanism of initiation of polymerization.

From *o*-phthalic acid, both a polymeric and a monomeric peroxide can be prepared:

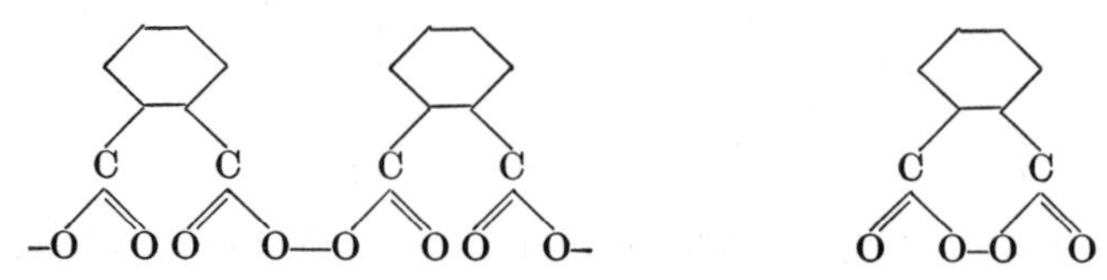

The polymeric peroxide is insoluble in styrene but it initiates polymerization; as reaction proceeds, the molecular weight of the peroxide gradually falls and it dissolves (Shah, Leonard and Tobolsky, 1951). The peroxidic linkages in the peroxide are more or less independent of one another and break at random so that monoradicals are produced. The end-groups of the polymers contain peroxidic groups which may later decompose and growth of the polymer chain may be resumed. The monomeric peroxide is a source of diradicals

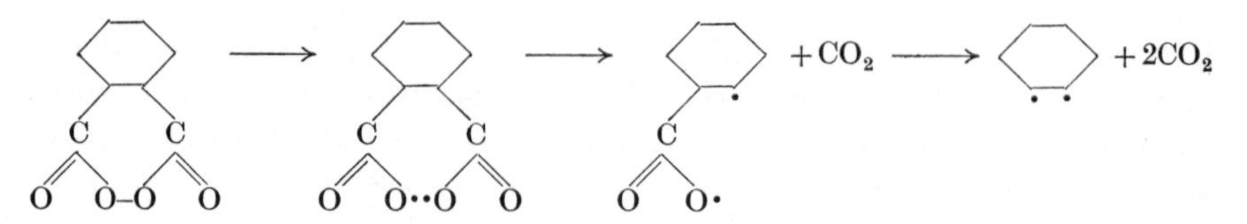

but it is very susceptible to induced decomposition, and there are large differences between the rates of decomposition in various solvents

(Russell, 1955; Greene, 1956). It is an inefficient initiator of polymerizations largely because of the self-termination of the polymer diradicals after the addition of a small number of monomer units.

Dibenzoyl peroxide may be used in redox systems; with the ferrous ion, the radical-producing reaction is probably

$$(C_6H_5.CO.O)_2+Fe^{++} \longrightarrow C_6H_5.CO.O\cdot+C_6H_5.CO.O^-+Fe^{+++} \quad (10)$$

Most of the reducing agents effective for this purpose are water-soluble although the peroxide itself is oil-soluble. Thorough study of these systems is complicated and (10) may take place at the interface between the two phases.

Decomposition of the di-aroyl peroxides can be promoted by *tert.*-amines; the important work of Horner has been summarized by Walling (1957). Radical polymerizations can be initiated at temperatures as low as 0°C but only a small fraction of the decomposed peroxide is utilized in this way. At higher temperatures, the rate of polymerization falls off quite rapidly because of consumption of the peroxide and amine. The first stage in the reaction is probably the formation of a polar intermediate which may subsequently decompose in several ways, including

$$C_6H_5.N(CH_3)_2+(C_6H_5.CO.O)_2 \longrightarrow [C_6H_5.N(CH_3)_2.O.CO.C_6H_5]^+[C_6H_5.CO.O]^-$$
$$\longrightarrow [C_6H_5.\dot{N}(CH_3)_2]^+ + C_6H_5.CO.O^- + $$
$$+C_6H_5.CO.O\cdot \quad (11)$$

The overall reaction (11) resembles (10) in that there is a one-electron transfer and scission of the peroxidic linkage. Experiments with dimethylaniline and a labelled di-aroyl peroxide have shown that both aroyloxy and aryl end-groups are introduced into polymers (Bevington and Lewis, 1960*b*); this result is in keeping with the fact that the aroyloxy radical produced in (11) may dissociate before it can engage in reaction with monomer.

The decompositions of di-aroyl peroxides can be accelerated by strong acids (Bartlett and Leffler, 1950); the catalysed reaction is probably non-radical in character. For *p*-methoxy-*p'*-nitrobenzoyl peroxide, the importance of the non-radical decomposition is increased by working with a polar solvent (Leffler, 1950); this may account for this peroxide not initiating the polymerization of polar monomers such as acrylonitrile. The decomposition of dibenzoyl peroxide in dimethyl formamide is catalysed by the chloride ion but the system fails to initiate polymerization (Bamford and White, 1960). The first step in this induced decomposition is believed to be the formation of benzoyl hypochlorite

$$(C_6H_5.CO.O)_2+Cl^- \longrightarrow C_6H_5.CO.O.Cl+C_6H_5.CO.O^-$$

Under controlled conditions, di-acetyl peroxide dissociates to radicals at about the same rate as dibenzoyl peroxide, but it is very susceptible to explosive decomposition even at room temperature. Other aliphatic di-acyl peroxides are less dangerous and dilauroyl peroxide is used commercially.

Although di-acetyl peroxide probably dissociates to acetyloxy radicals, these are so unstable that yields of carbon dioxide are almost quantitative even when iodine is present (Rembaum and Szwarc, 1955*a*). The dissociation of the peroxide is therefore effectively

$$(CH_3.CO.O)_2 \longrightarrow 2CH_3\cdot + 2CO_2$$

and the compound is commonly used as a source of methyl radicals in the region of 60°C. Similarly, dipropionyl peroxide has been used to generate ethyl radicals (Smid and Szwarc, 1956). The rates of decomposition of

TABLE 2.2

DECOMPOSITIONS OF GASEOUS DI-ACYL PEROXIDES

Peroxide	A (sec^{-1})	E (kcal/mole)	Reference
di-acetyl	$1{\cdot}8 \times 10^{14}$	29·5	Rembaum and Szwarc (1954)
dipropionyl	$2{\cdot}5 \times 10^{14}$	30	Rembaum and Szwarc (1955*a*)
dibutyryl	$1{\cdot}9 \times 10^{14}$	29·6	Rembaum and Szwarc (1955*a*)

gaseous di-acetyl, dipropionyl and dibutyryl peroxides are very similar (see Table 2.2); decompositions in non-polar solvents proceed at about the same rates if [peroxide] is low, but in alcohols there are induced reactions.

Peresters

Peresters are useful sources of alkoxy and acyloxy radicals

$$R.CO.O.O.R' \longrightarrow R.CO.O\cdot + R'.O\cdot \qquad (12)$$

the rate of decomposition usually being intermediate between those of the appropriate di-alkyl and di-acyl peroxides; thus *tert.*butyl perbenzoate can act as a source of *tert.*butoxy radicals at temperatures rather lower than those at which di-*tert.*butyl peroxide can be used conveniently, and as a source of benzoyloxy radicals at temperatures at which dibenzoyl peroxide decomposes rather rapidly.

The alkoxy and acyloxy radicals may dissociate further, as discussed already. It has been suggested (see, for example, Bartlett and Hiatt,

1958) that primary dissociation may involve a concerted break of oxygen–oxygen and carbon–carbon bonds. In the case of the *tert.*butyl compounds, the evidence for the single-stage reaction

$$R.CO.O.O.C_4H_9 \longrightarrow R\cdot + CO_2 + C_4H_9.O\cdot \qquad (13)$$

is based on a relationship between the activation energy for dissociation of the perester and the resonance stabilization of the radical R•. If the primary dissociation is according to (12), stretching of the oxygen–oxygen bond is accompanied by an increase of the energy of the molecule until it reaches a steady level corresponding to the separated acyloxy and alkoxy radicals (see Fig. 2.3). Subsequent dissociation of the acyloxy radical causes the energy to drop to the level (*a*) if R• is moderately stable, or to (*b*) if it is highly stabilized. It was suggested that the dotted extensions to the lines labelled (*a*) and (*b*) would correspond to the changes in energy with bond stretching for systems in which (13) occurs. In this case, the activation energy for dissociation would depend upon the properties of the radical R•.

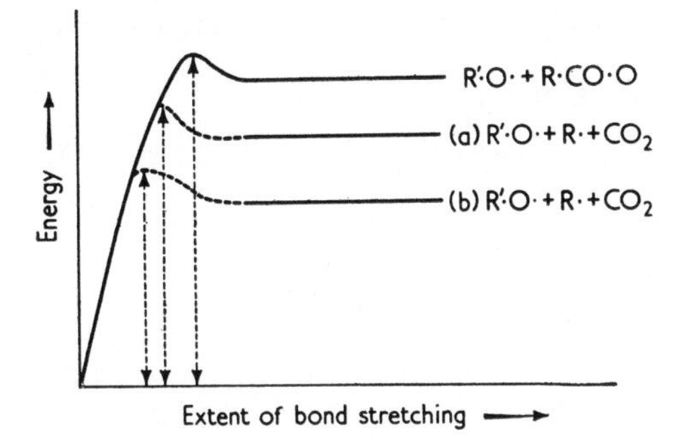

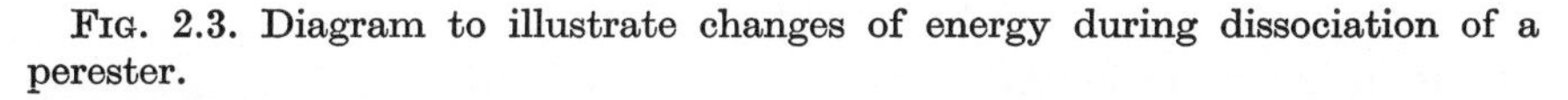

FIG. 2.3. Diagram to illustrate changes of energy during dissociation of a perester.

Very high yields of carbon dioxide were found for decompositions of these peresters in inert solvents. This is so also for dibenzoyl peroxide and is not proof that (13) occurs. The crucial test would be to determine the carbon dioxide produced during decompositions in the presence of radical scavengers. If (13) occurs exclusively, production of carbon dioxide would be unaffected by the scavenger, but if primary dissociation is by (12) the yield of carbon dioxide should tend to zero as [scavenger] is raised.

Other Peroxides

Several peroxydicarbonates, R.O.CO.O.O.CO.O.R, have been described (McBay and Tucker, 1954; McBay, Tucker and Milligan, 1954) and used to a limited extent as sources of radicals in solution. They decompose at quite low temperatures and are somewhat hazardous. The mono-alkyl carbonate radicals first generated may subsequently lose carbon dioxide:

$$(R.O.CO.O)_2 \longrightarrow 2R.O.CO.O\cdot \longrightarrow 2R.O\cdot + 2CO_2$$

and the alkoxy radicals so formed may decompose further. When the di-*iso*propyl compound is used to initiate the polymerization of styrene (Cohen and Sparrow, 1950), yields of carbon dioxide are depressed because primary radicals are captured, thus:

$$R.O.CO.O\cdot + CH_2{:}CH.C_6H_5 \longrightarrow R.O.CO.O.CH_2.CH(C_6H_5)\cdot \qquad (14)$$

Chemical tests on the polymer confirm that it contains end-groups of the type required by (14).

The *tert.*butyl ester of N-phenylperoxycarbamic acid decomposes by a first-order process (O'Brien, Beringer and Mesrobian, 1957). The primary dissociation is probably

$$C_6H_5.NH.CO.O.O.C_4H_9 \longrightarrow C_6H_5.NH.CO.O\cdot + C_4H_9.O\cdot$$

but the further dissociation

$$C_6H_5.NH.CO.O\cdot \longrightarrow C_6H_5.NH\cdot + CO_2$$

evidently occurs very readily since even in styrene the yield of carbon dioxide is as high as 85% of the maximum.

The trans-annular peroxide ascaridole

$$CH_3\text{—}C\begin{matrix}\diagup CH{=}CH \diagdown \\ \text{—O————O—} \\ \diagdown CH_2\text{–}CH_2 \diagup\end{matrix}C\text{—}CH(CH_3)_2$$

and its dihydro derivative have been used to initiate polymerizations, for example by Zand and Mesrobian (1955). The reactions have the characteristics of mono-radical polymerizations, although diradicals would be expected by scission of the O–O bond. The peroxides are quite efficient initiators so that ring-closure of very small polymer diradicals cannot be as important as it is when other sources of diradicals are decomposed in the presence of monomers. It is supposed that the diradicals first formed readily change into mono-radicals, perhaps by reactions such as

$$\begin{matrix}\cdot O \\ CH_3\end{matrix}\!\!>\!C\!<\!\!\begin{matrix}CH_2\text{—}CH_2 \\ CH_2\text{—}CH_2\end{matrix}\!\!>\!C\!<\!\!\begin{matrix}O\cdot \\ CH(CH_3)_2\end{matrix} \longrightarrow \begin{matrix}\cdot O \\ CH_3\end{matrix}\!\!>\!C\!<\!\!\begin{matrix}CH_2\text{—}CH_2 \\ CH_2\text{—}CH_2\end{matrix}\!\!>\!C{=}O + (CH_3)_2CH\cdot$$

and that these mono-radicals are the initiating species.

C. Azo Compounds

Azo compounds can dissociate into radicals, generally according to the equation

$$R.N{:}N.R \longrightarrow 2R\cdot + N_2 \qquad (15)$$

and in only a few cases is there any indication that the radical $R.N{:}N\cdot$ is first formed and subsequently dissociates. An important feature of a direct decomposition according to (15) is that recombination of the radicals gives a product distinct from the starting material.

If $R\cdot$ is an aliphatic hydrocarbon radical, (15) occurs only at fairly high temperatures; thus, azo*iso*propane is too stable for use as a thermal sensitizer for radical polymerizations, but it is a useful photosensitizer although susceptible to transfer reactions. If R is aromatic, E_{15} may be low enough for the compound to be a thermal sensitizer, as shown by data in Table 2.3 (Cohen and Wang, 1955). The reduction in E_{15} accompanying substitution in the methyl group is attributed to resonance stabilization of the radicals. In some cases, e.g. the diphenylmethyl radical, the stabilization may be so great that the radical is comparatively unreactive.

Table 2.3

DISSOCIATIONS OF AZO DERIVATIVES OF HYDROCARBONS

Group R	Activation energy (kcal/mole)	Entropy of activation (cal/mole/deg.)	Temp. (°C) at which $k = 10^{-5}$ sec^{-1}
CH_3	50·2	11	252
$CH(CH_3)_2$	40·9	1	203
$CH(CH_3)(C_6H_5)$	32·6	7	86
$CH(C_6H_5)_2$	26·6	2	36

Azo*iso*butyronitrile, $(CH_3)_2C(CN).N{:}N.C(CH_3)_2.CN$, is the best known of the azo initiators. Its thermal decomposition in solution is accurately first order, and there are only small differences between the rates in various solvents. The compound is not susceptible to attack by radicals so that induced decomposition and transfer reactions are unimportant. It also dissociates to free radicals under the influence of near u.v. light, and can be used as a photosensitizer of polymerizations.

The decomposition of azo*iso*butyronitrile is usually written as

$$(CH_3)_2C(CN).N{:}N.C(CH_3)_2.CN \longrightarrow 2(CH_3)_2C(CN)\cdot + N_2$$

but some of the radicals react as $(CH_3)_2C{:}C{:}N\cdot$ (Talât-Erben and Bywater, 1955*b*). In suitable systems, appreciable quantities of dimethylketene–cyan*iso*propylimine are formed by the combination of $(CH_3)_2C(CN)\cdot$ and $(CH_3)_2C{:}C{:}N\cdot$ radicals, and accompany the tetramethylsuccinodinitrile formed by combination of pairs of radicals reacting in the first form. The spatial arrangements of the atomic nuclei in the two radicals are the same, and evidently the electron-density at the nitrogen atom is sufficient for an appreciable fraction of the radicals to react in the abnormal form. During photolysis of azo*iso*butyronitrile at 25°C, the molar conversion to the ketene-imine is about 58% (Smith and Rosenberg, 1959). The ketene-imine is not stable and at about 60°C in inert solvents it rearranges to tetramethylsuccinodinitrile (Talât-Erben and

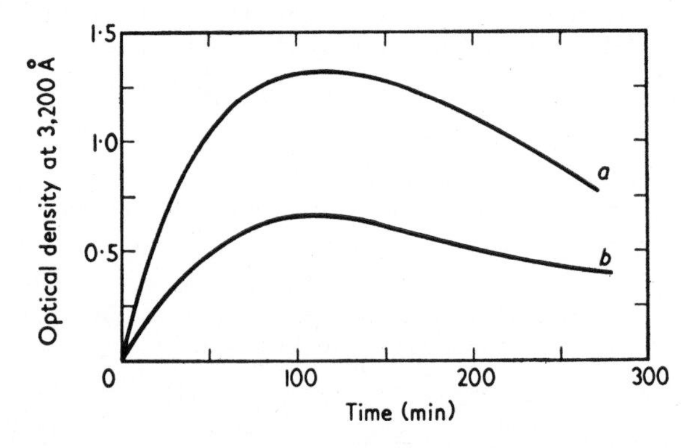

FIG. 2.4. Curves to show changes of [ketene-imine] with time during decomposition of azo*iso*butyronitrile in toluene at 80·4°C. Initial values of [azo*iso*butyronitrile]: *a*—0·141 mole/l.; *b*—0·072 mole/l. (After Talât-Erben and Bywater, 1955*b*.)

Isfendiyaroğlu, 1958, 1959; Smith and Carbone, 1959). The ketene-imine initiates polymerizations with an efficiency approaching that of the parent azonitrile (Hammond, Trapp, Keys and Neff, 1959).

The ketene-imine can be determined by measurements of the optical density of the solution at 3200 Å. During decomposition of the azonitrile, [ketene-imine] builds up to a maximum and then decays (see Fig. 2.4). This probably accounts for the observation (Bevington, 1954) that *iso*-butyronitrile is apparently formed in substantial quantities in the early stages of the decomposition of azo*iso*butyronitrile but later is converted into tetramethylsuccinodinitrile. It now seems that most of the mononitrile is produced not by disproportionation of 2-cyano-2-propyl radicals, but rather by decomposition of the ketene-imine during the analysis; this decomposition is responsible for the apparent disappearance of *iso*butyronitrile during the later stages of the reaction.

The possibility of the ketene-imine being a source of radicals introduces a complication into the chemistry of the azonitrile. In the later stages of the decomposition of the azo compound, the rate of generation of radicals may be higher than expected owing to a significant contribution from the decomposition of the ketene-imine produced early in the decomposition; no abnormality would be expected in the rate of production of nitrogen. Another complication arises from the fact that 2-cyano-2-propyl radicals react readily with oxygen

$$(CH_3)_2C(CN)\cdot + O_2 \longrightarrow (CH_3)_2C(CN).O.O\cdot$$

and the reactivity of the peroxy radical is, in certain reactions at least, quite different from that of the original radical. Combination of the peroxy radicals with 2-cyano-2-propyl radicals gives a peroxide which may subsequently dissociate to radicals. Talât-Erben and Önol (1960) have shown that when 2-cyano-2-propyl radicals are generated in the presence of oxygen, there may be appreciable yields of the hydroperoxide $(CH_3)_2C(CN).O.O.H$; this product is presumably formed by the abstraction of hydrogen from the solvent by the peroxy radical. The hydroperoxide can initiate polymerization. The alkoxy radical formed from the substituted di-alkyl peroxide or from the hydroperoxide may dissociate to acetone and the radical ·CN. These peroxides could cause discrepancies between the rate of production of radicals and the rate of evolution of nitrogen in the later stages of the decomposition of the azonitrile.

Van Hook and Tobolsky (1958) concluded that the velocity constant for the decomposition of azo*iso*butyronitrile over the range 37° to 100°C is best expressed as

$$k = 1{\cdot}58.10^{15} \exp(-30{\cdot}8\,\text{kcal}/RT)\,\text{sec}^{-1}$$

It is conveniently used as a thermal sensitizer for radical polymerizations at temperatures between 45° and 65°C. The rate of decomposition in solution, deduced from the rate of consumption of the radical scavenger diphenylpicrylhydrazyl (Bawn and Mellish, 1951), refers only to the production of *available* radicals, and wastage by some form of cage recombination is not detected (Walling, 1954). The rate of decomposition is decreased by applying high pressures and there are indications that the importance of the cage-effect also may be increased (Ewald, 1956).

The general features of the decomposition of azo*iso*butyronitrile would probably be found for other azonitriles and compounds such as

azo-esters. Certain azonitriles such as

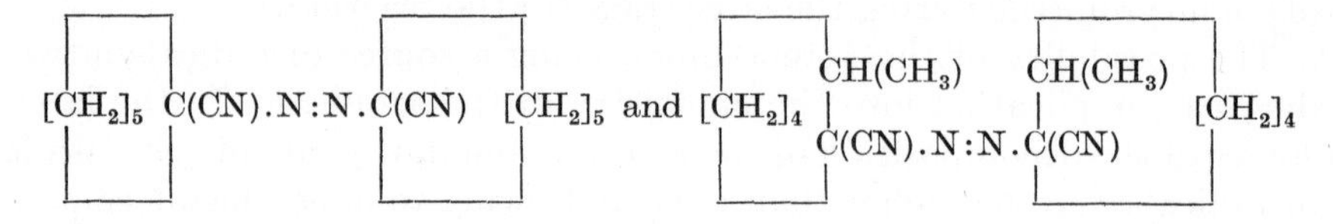

in which the azo grouping links two ring systems, are important, being fairly stable thermally (Overberger, Biletch, Finestone, Lilker and Herbert, 1953). Dimethylazo*iso*butyrate decomposes according to a first-order law with

$$k = 1\cdot31.10^{-15}\exp(-31\cdot1\,\text{kcal}/RT)\,\text{sec}^{-1}$$

It would be expected to give 2-carbomethoxy-2-propyl radicals, $(CH_3)_2C(COOCH_3)\cdot$, and the products of its decomposition in solution are in accord with this (Bickel and Waters, 1950*a*). Spectroscopic evidence shows, however, that there is also an unstable product (MacKie and Bywater, 1957) which may well result from some of the radicals reacting in the form $(CH_3)_2C{:}C(OCH_3).O\cdot$.

Another type of azonitrile, useful for special applications, contains additional functional groups; thus, the azonitrile prepared from a keto acid, such as laevulinic acid, is water-soluble because of its carboxylic acid groups. Polymers prepared using such initiators have reactive end-groups through which polymer molecules may be linked together. Products of the ketene-imine type are formed during the decompositions of some of these azonitriles (Haines and Waters, 1958).

Overberger and Lapkin (1955) found that the compound

$$\begin{array}{c} C_6H_5.\overset{}{C}H.N{:}N.CH.C_6H_5 \\ \quad| \qquad\qquad | \\ [CH_2]_8 \qquad [CH_2]_8 \\ \quad| \qquad\qquad | \\ C_6H_5.CH.N{:}N.CH.C_6H_5 \end{array}$$

does not initiate the polymerization of styrene although it apparently dissociates quite readily to diradicals. It appears that the radicals may take up a few molecules of monomer, but that termination by ring-closure occurs very readily. Similar conclusions have been reached for the diradicals formed from other cyclic initiators.

Azo compounds of other types also have been mentioned as initiators of radical polymerizations. Azodibenzoyl, $C_6H_5.CO.N{:}N.CO.C_6H_5$, might be a useful source of benzoyl radicals (Leffler and Bond, 1956) which may dissociate further to phenyl radicals and carbon monoxide. The diazothio-ethers, R.N:N.S.R′, where R and R′ are aromatic groups, function both as initiators and transfer agents (Reynolds and

Cotten, 1950). Diazo-aminobenzene initiates the polymerization of styrene at moderate temperatures, but plots of rate of polymerization against [initiator] show maxima (Haward and Simpson, 1951) indicating that the compound can function also as a retarder.

The N-nitroso-anilides, R′N(NO).CO.R″, where R′ is an aryl group and R″ may be aryl or alkyl, have been used as thermal sensitizers for radical polymerizations at temperatures as low as 0°C. The rate-determining step in the decomposition of N-nitroso-acetanilide is the rearrangement to the diazo-acetate which subsequently dissociates (Huisgen and Horeld, 1949):

$$C_6H_5.N(NO).CO.CH_3 \longrightarrow C_6H_5.N{:}N.O.CO.CH_3$$
$$\longrightarrow C_6H_5\cdot + N_2 + CH_3.CO.O\cdot$$

Decomposition is, however, not a simple process; it is frequently accompanied by development of colour and the yields of nitrogen may be only 80% of theoretical. Analyses of polymers for fragments derived from N-nitroso-anilides have been performed using materials containing bromine (Hey and Misra, 1947) or labelled with carbon-14 (Bevington and Lewis, 1960*b*). It is agreed that the aryl radical, R′, initiates polymerization but there are conflicting results concerning the radical R″.

D. Various other Sources of Radicals

Compounds of many other types can act as sources of radicals, some of them at relatively low temperatures; an extensive account has been given by Walling (1957). Some of these compounds have been used to initiate radical polymerizations but in many cases there have been only preliminary investigations. In this section, a few initiating systems, showing features of special interest and importance, are considered.

Certain mono- and disulphides, R.S.R and R.S.S.R, dissociate to radicals either photochemically or thermally at moderate temperatures, and some are used also in redox reactions such as

$$Fe^{++} + R.S.S.R \longrightarrow Fe^{+++} + RS^- + RS\cdot$$

Many of these compounds are best known as cross-linking agents for rubber, but some have been used also as initiators of polymerizations; in some cases, complications arise from transfer to initiator which may be accompanied by retardation.

When tetramethylthiuram disulphide, $(CH_3)_2N.CS.S.S.CS.N(CH_3)_2$, is used to promote vulcanization, temperatures greater than 100°C are usually employed; it is an initiator for the polymerization of the more reactive monomers, such as styrene, at temperatures as low as 60°C (Otsu, Nayatani, Muto and Imai, 1958). The rate of dissociation has been measured using diphenylpicrylhydrazyl (Ferington and Tobolsky, 1955). Complicated relationships between [initiator] and the rate of polymerization are due to it acting also as a retarder (Otsu and Nayatani, 1958). It is quite likely that the primary dissociation of tetra-alkylthiuram disulphides may involve scission of a carbon–sulphur or of a sulphur–sulphur bond (Dogadkin and Shernshnev, 1959); in the latter case, the resulting radicals may undergo further dissociation:

$$(R_2N.CS.S)_2 \longrightarrow 2R_2N.CS.S\cdot \longrightarrow 2R_2N\cdot + 2CS_2$$

Polystyrene prepared using an initiator of this type is a photo-sensitizer for polymerizations, so evidently it contains labile end-groups, considered to be $R_2N.CS.S\cdot$ (Otsu, 1957). Ferington and Tobolsky (1958) also examined the effects upon polymerizations of tetramethylthiuram monosulphide, a corresponding tetrasulphide and diphenyl disulphide; the chief significance of their results is in connection with the mechanism of transfer to initiator.

A cyclic disulphide has been examined (Russell and Tobolsky, 1954) as source of diradicals by the photo-dissociation

$$O\left\langle\begin{matrix} CH_2.CH_2.S \\ | \\ CH_2.CH_2.S \end{matrix}\right. \longrightarrow \cdot S.CH_2.CH_2.O.CH_2.CH_2.S\cdot$$

Just like other cyclic compounds which can give diradicals, this substance is not an initiator of polymerizations; it is supposed, as in other cases, that termination by ring closure occurs at a very early stage in the reaction.

Photolysis of hydrazine gives $\cdot NH_2$ radicals which can initiate polymerizations, but wavelengths less than 2600 Å are required (Uri, 1952). The $\cdot NH_2$ radical can be generated from hydroxylamine by an electron-transfer reaction with, for example, the titanous ion

$$M^{n+} + NH_2.OH \longrightarrow M^{(n+1)+} + \cdot NH_2 + OH^-$$

and polymerizations can be initiated in aqueous acidic media (Davis, Evans and Higginson, 1951). If the $\cdot NH_2$ radical escapes capture by monomer, it may oxidize another metal ion

$$M^{n+} + \cdot NH_2 \longrightarrow M^{(n+1)+} + NH_2^- \qquad (16)$$

the amide ion being later converted to ammonia. Competition between the initiation process and (16) is rather similar to that discussed in connection with the hydroxyl and other radicals; it could be used for comparison of the reactivities of monomers towards the $\cdot NH_2$ radical.

Redox systems for initiating polymerizations can be formed from various inorganic oxidizing agents, e.g. ceric salts, permanganates, chlorites and hypochlorites, with reducing agents both organic and inorganic; such systems have been described by Bovey, Kolthoff, Medalia and Meehan (1955), but in many cases the detailed chemistry is unknown. Certain much simpler electron-transfer reactions should be mentioned. The decomposition of water by the cobaltic ion initiates vinyl polymerizations (Bawn and White, 1951) through the agency of the hydroxyl radical

$$Co^{3+}OH^- \longrightarrow Co^{2+} + \cdot OH \qquad (17)$$

Radicals can also be formed by similar transfers between the cobaltic ion and organic molecules, e.g.

$$H.COOH + Co^{3+} \longrightarrow H.CO.O\cdot + H^+ + Co^{2+}$$
$$H.COO^- + Co^{3+} \longrightarrow H.CO.O\cdot + Co^{2+}$$
$$CH_3OH + Co^{3+} \longrightarrow CH_3.O\cdot + H^+ + Co^{2+}$$

Complexes of the ferric ion with anions such as OH^-, Cl^- and CNS^-, do not form radicals by thermal reaction similar to (17) at convenient temperatures, but photochemical processes may occur (Dainton and James, 1958) and form useful means for generating radicals in aqueous media:

$$Fe^{3+}X^- \xrightarrow{h\nu} Fe^{2+} + X\cdot \qquad (18)$$

If all the radicals produced in (18) are captured by monomer, the rate of initiation might be put equal to the rate of appearance of ferrous ions in the system, but these ions may be formed also in termination reactions (Dainton and Tordoff, 1957) (see Chapter 7, C). Also if [monomer] is very low, the reverse of (18) may become significant. Photochemical electron-transfers of the general forms

$$M^{z+}.H_2O \longrightarrow M^{(z+1)+} + OH^- + H\cdot$$
$$A^{z-}.H_2O \longrightarrow A^{(z-1)-} + OH^- + H\cdot$$
$$A^{z-}.H_3O^+ \longrightarrow A^{(z-1)-} + H_2O + H\cdot$$

have been considered in detail (Dainton and James, 1958) and references to previous work on the subject have been given. The hydrogen atoms can initiate polymerizations but there are many complications, including, in some cases, those arising from formation of complexes between the monomer and the ion.

The co-polymerization of styrene and methyl methacrylate can be initiated by metallic lithium; unlike polymerizations initiated by *n.*-butyl lithium or by metallic sodium or potassium, this reaction has some of the characteristics of a radical process. It is believed that an electron is transferred from an atom of the metal to a molecule of monomer to give a radical-ion, $\cdot CH_2.CXY^-Li^+$, which can act as a centre for both radical and anionic polymerizations (O'Driscoll and Tobolsky, 1958). A random co-polymer results from the radical reaction and almost pure polymethyl methacrylate from the ionic polymerization. The relative importances of the two types of polymerization can be altered by changing the nature of the solvent; the radical reaction requires the higher energy of activation and so it becomes less important at low temperatures.

A radical-ion is produced also by electron-transfer from the sodium-naphthalene complex to a hydrocarbon monomer such as styrene; it may initiate both radical and cationic polymerizations (Szwarc, Levy and Milkovich, 1956). If the rate of initiation is high, the growing radicals are soon terminated by mutual interaction; if the system is completely free from impurities, however, the ionic polymerization can continue until the monomer is consumed, and even then the centres are still present. Szwarc has described these materials as "living polymers", and has stressed their value for the production of mono-disperse polymers and of block co-polymers.

Photo-initiation of radical polymerizations is of special significance in connection with the determination of the velocity constants for the separate steps; the production of radicals can be started and stopped abruptly, and this is required for studies of the non-stationary states in radical polymerizations (Bamford, Barb, Jenkins and Onyon, 1958). Another important feature is that the rate of production of radicals is independent of temperature. Photo-initiation is of particular importance for polymerizations at, or below, room temperature; the generation of radicals by thermal reactions is not suitable in these cases since it is necessary either to preserve and then to dispense an unstable substance at very low temperatures, or to use a multi-component initiating system, one ingredient being added to the reaction mixture at the very last stage. The rate of production of radicals from a photosensitizer can be adjusted by altering either [sensitizer] or the intensity of the light, a fact utilized in a method for studying transfer to initiator (see Chapter 5 D). It is desirable that the photochemically active light should be in the near u.v. to simplify experimental arrangements and to avoid production of radicals directly from the monomer. The extinction coefficient for the sensitizer should be small so that there is uniform absorption of light and production of radicals throughout the reaction mixture. A practical difficulty concerns the exact positioning of lamp and reaction vessel necessary to ensure reproducibility.

At various points in the preceding sections, mention has been made of photosensitizers; most of them were substances which can also give radicals by thermal dissociation at moderate temperatures. It is necessary also to consider a few examples of photosensitizers which are too stable to act as thermal sensitizers for polymerizations.

Certain dyes have been used to initiate polymerizations, as for example in the work of Oster (1954) with preparations of chlorophyll, but the precise natures of the initiating radicals are unknown. Metal alkyls are frequently employed as sources of radicals by photolysis; usually they are used in the gas phase, but lead tetra-ethyl, for example, has been mentioned as a photo-initiator of polymerizations in the liquid phase (Marvel and Woolford, 1958).

Carbonyl compounds are very important as sources of radicals by photo-dissociation. The simpler aldehydes and ketones require rather short wavelengths for dissociation, and not all the photolyses are simple processes; thus, production of radicals from aliphatic methyl ketones

$$R.CO.CH_3 \longrightarrow R\cdot + CH_3.CO\cdot$$

may be accompanied by intramolecular rearrangement to acetone and an olefin (Nicholson, 1954). Diketones absorb at longer wavelengths and

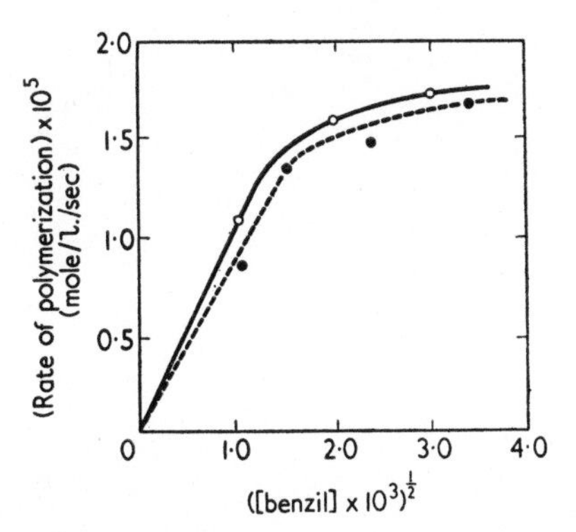

Fig. 2.5. Effect of [benzil] upon rate of photo-sensitized polymerization at 25°C. —○— styrene; ——●—— methyl methacrylate.

are more suitable as photosensitizers for polymerizations. The primary dissociations give acyl radicals which may lose carbon monoxide before engaging in reaction

$$R.CO.CO.R \longrightarrow 2R.CO\cdot$$
$$R.CO\cdot \longrightarrow R\cdot + CO$$

Benzoin is a photo-initiator for radical polymerizations, and is believed to dissociate thus:

$$C_6H_5.CO.CH(OH).C_6H_5 \longrightarrow C_6H_5.CO\cdot + C_6H_5.CH(OH)\cdot$$

It is more efficient than benzil as a photo-initiator of the polymerization of methyl methacrylate at 25°C (Chinmayanandam and Melville, 1954); the rate of initiation by benzoin is about twice that by benzil at the same molar concentration and light intensity, in spite of the fact that benzil shows the stronger absorption. The high extinction coefficient for benzil can lead to non-uniform production of radicals and deviations from the relationship

$$\text{rate of polymerization} \propto [\text{benzil}]^{1/2}$$

as [benzil] is increased while the intensity of the light remains constant (see Fig. 2.5).

Although simple ethers of benzoin are photosensitizers, derivatives such as the oxime are not effective (Mochel, Crandall and Peterson, 1955); it seems, therefore, that the carbonyl group is the site of the photochemical activity. More detailed examination of benzoin and its methyl ether has revealed that polymers contain considerably more combined initiator than can be accounted by the initiation process; a likely explanation is discussed in Section B.*1* of Chapter 4.

Chapter 3

Initiation of Polymerization

A. Introduction

In this chapter, the reactions of small radicals with monomers are considered; these reactions initiate the growth of polymer chains. Section B is devoted mainly to the significance of the efficiency of initiation, i.e. that fraction of the radicals generated in a system which actually initiates polymerization chains. The efficiency of initiation is closely related to the rate of initiation, and in Section C the important methods for measuring this rate are discussed critically. Reliable determination of rates of initiation is essential for kinetic studies of radical polymerizations designed for evaluating the velocity constants for the various component reactions.

The last section of this chapter is concerned with methods for comparing the reactivities of monomers towards various primary radicals; there may be considerable differences between these reactivities. The reasons for these differences are best discussed when considering the factors which govern the rates of the various growth reactions in polymerizations (see Chapter 4, B.*3*).

Throughout this chapter, much emphasis is placed on the use of isotopically labelled initiators. Their great value lies in their use for very sensitive, accurate and specific analyses for fragments of initiator incorporated in polymer. These analyses can lead to information concerning the mechanisms and rates of the initiation process.

B. Rates and Efficiencies of Initiation

There are several distinct methods for determining rates of initiation in radical polymerizations but no single one is applicable to all cases. Each procedure depends upon assumptions, in some cases unproved and so drastic as to make the methods very uncertain. Some of the methods involve no measurements on the polymer formed during the reaction, but others depend upon analysis or characterization of the polymer.

If the rate of initiation of polymerization and the rate of decomposition of the initiator are known, the efficiency of initiation can be calculated from

$$\text{efficiency} = \frac{\text{rate of initiation}}{n \, . \, k_d[\text{initiator}]}$$

where k_d is the velocity constant for decomposition of the initiator, and n, the number of radicals produced from each molecule of initiator, is normally 2 but may be 1 in certain cases. The velocity constant k_d refers to the primary dissociation of initiator and not to the total rate of consumption, which may include contributions from induced reactions. Efficiencies of less than 100% can be caused in two ways; the initiator may, in part, decompose by a mechanism which does not produce radicals capable of initiating polymerization, or, alternatively, some of the radicals may not initiate reaction chains but instead engage in other reactions. The wastage reactions might be:

(*a*) a direct interaction of radicals derived from the initiator;

(*b*) a termination process involving a polymer radical and an initiator radical, i.e. *primary radical termination*;

(*c*) reaction of the radical with another component of the system.

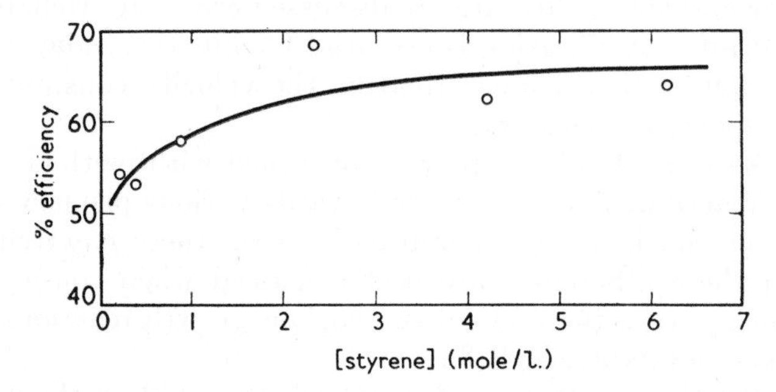

FIG. 3.1. Efficiency of initiation by azo*iso*butyronitrile in polymerization of styrene at 60°C.

Measurement of rates of initiation by azo*iso*butyronitrile by various techniques leads to efficiencies of only about 60% over a wide range of concentrations of monomer (see Fig. 3.1). Calculations of the relative rates of the initiation reaction and those by which radicals may be wasted, assuming reasonable values for the velocity constants and the concentrations of radicals, show that initiation should be overwhelmingly favoured unless [monomer] is very low. The comparatively low efficiency is commonly explained in terms of the cage-effect; the solvent molecules are considered as forming a barrier which prevents the separation of radicals and so encourages their interaction. Noyes (1955) has treated the problem from rather a different point of view, pointing out that ordinary kinetic principles cannot be applied to the competition between

$$2R\cdot \longrightarrow R_2 \tag{1}$$

and

$$R\cdot + M \longrightarrow R.M\cdot \tag{2}$$

because the distribution of R· radicals is not random. The radicals are generated in pairs, and so, until they have diffused apart, the wastage reaction is favoured. Three distinct types of recombination have been considered. *Primary recombination* occurs within about 10^{-11} sec of the generation of the radicals and before they have separated by more than a molecular diameter. *Secondary recombination* occurs within about 10^{-9} sec of the dissociation, so that although the radicals have undergone several diffusive displacements there is a finite probability of re-encounter. *Recombination in the body of the solution* takes place between radicals which have escaped primary and secondary recombination and also reaction with a scavenger; this reaction most probably involves radicals originating from different molecules of initiator.

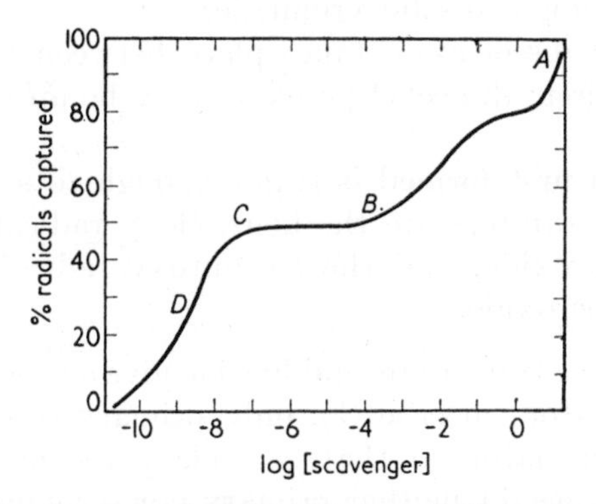

FIG. 3.2. Sketch showing expected dependence of efficiency of a scavenger upon its concentration. (After Noyes, 1955.)

Primary and secondary recombination together may be referred to as cage-recombination, but Noyes prefers the term *geminate recombination* to suggest the interaction of radicals formed from the same molecule of initiator. The fraction of radicals captured by a very reactive scavenger would vary with its concentration as indicated in Fig. 3.2. Probably no scavenger can suppress primary recombination and only those considerably more reactive than any ordinary monomer can interfere with secondary recombination. The part *AB* of the curve in Fig. 3.2, therefore, cannot be achieved when the scavenger is a monomer. In the region *BC*, radicals either react with scavenger or undergo geminate recombination; it is only when the concentration of scavenger is very low that there is recombination of radicals in the body of the solution. The curve in Fig. 3.1 corresponds, therefore, to *BCD* in Fig. 3.2.

If the recombination (1) regenerates a molecule of the original initiator, it is almost impossible to prove that the reaction has occurred; the efficiency of initiation would appear to be 100% over a range of concentrations of monomer, and only at very low concentrations would there be any significant drop. The only way to detect geminate recombination for such a system would require the use of a scavenger so reactive that it could interfere with this type of recombination; for a given concentration of initiator, the rate of consumption of this scavenger would then be greater than the rate of initiation of polymerization.

Many initiators dissociate to radicals which may themselves dissociate later giving smaller radicals; for initiators of this type, recombination of radicals may give products distinct from the original molecule. Three cases must be considered:

(*a*) the two dissociations occur simultaneously or within 10^{-11} sec, an example being azo*iso*butyronitrile;
(*b*) the second dissociation takes place between 10^{-11} and 10^{-9} sec after the first; di-acetyl peroxide may be an example of such a system;
(*c*) the radical first formed is stable enough to survive for at least 10^{-9} sec; examples are the benzoyloxy radical formed from dibenzoyl peroxide, and the *tert.*butoxy radical formed from di-*tert.*butyl peroxide.

In case (*a*), all three types of recombination give a product different from the original initiator; in case (*b*), only primary recombination would regenerate the initiator so that secondary recombination could be detected, but in case (*c*), neither primary nor secondary recombination could be detected.

In the case of azo*iso*butyronitrile, the overall rate of decomposition can be measured by the nitrogen evolution. The low efficiencies of initiation have been confirmed (Hammond, Sen and Boozer, 1955; Bevington, 1955) by finding in the reaction mixture products which could only have been formed by the interaction of 2-cyano-2-propyl radicals. The quantities of these waste products, together with the initiator fragments which become incorporated in polymer, are equivalent to the nitrogen evolved. If [monomer] is in the region where efficiency of initiation is independent of concentration, the waste products are formed only as the result of geminate recombination; at very low concentrations of monomer, the yields of waste products are larger, recombination in the body of the solution being significant.

The ideas on the various types of recombination might be tested by experiments involving an unsymmetrical azo initiator. If the initiator gives radicals $R_a\cdot$ and $R_b\cdot$, geminate recombination would give the product $R_a{\cdot}R_b$; at low concentrations of monomer when recombination in the body of the solution becomes significant, the products $(R_a)_2$ and $(R_b)_2$ also should be formed. Determination of these products at very low concentrations in mixtures would be difficult experimentally, but it should be possible by using labelled initiator and the technique of isotope dilution analysis. Similar experiments might be performed with mixtures of azo initiators; in this case, the unsymmetrical waste product should appear only when [monomer] is very low.

When dibenzoyl peroxide is decomposed thermally in the presence of radical scavengers, the production of carbon dioxide can be suppressed completely (Barson and Bevington, 1958); if the same concentrations of scavengers are used with azo*iso*butyronitrile, however, geminate recombination of some 2-cyano-2-propyl radicals is not prevented. Evidently, therefore, in an inert solvent the benzoyloxy radicals survive on the average for more than 10^{-9} sec; it has been estimated (Bevington, and Toole, 1958) that at 60°C in such a solvent, the half-life is about 10^{-4} sec, decreasing to about 3×10^{-5} sec at 80°C. Primary or secondary recombination of radicals for this initiator must involve benzoyloxy radicals and so be undetectable. During photolysis of the peroxide, however, some carbon dioxide is produced even in the presence of very high concentrations of scavengers (Bevington and Lewis, 1958); some of the benzoyloxy radicals evidently dissociate within 10^{-9} sec of their generation, and so geminate recombination ought to be detectable. When dibenzoyl peroxide is used as a thermal sensitizer for polymerizations, the rate of initiation is almost independent of [monomer] over a wide range (Bevington, 1957), indicating that recombination of initiator radicals in the body of the solution can be neglected. Over this range of concentrations, the yields of carbon dioxide may vary quite considerably, showing that both benzoyloxy and phenyl radicals are present at isolated sites in the solution, and that recombination of initiator radicals in the body of the solution could be detected if it occurred, since it would give products, viz. phenyl benzoate and diphenyl, distinct from the original peroxide.

The acetyloxy radicals formed by dissociation of di-acetyl peroxide are very unstable and dissociate to methyl radicals. In solution, scavengers do not suppress the combination of these methyl radicals, indicating that it occurs as a geminate recombination (Rembaum and Szwarc, 1955*b*); in the gas phase, however, the combination can be prevented. This difference indicates that in solution the dissociation of the acetyloxy radicals occurs within about 10^{-9} sec of their generation. This peroxide has not been much studied as an initiator of polymerization, but it is anticipated that efficiencies of less than 100% would be found, since geminate recombination gives a product distinct from the original initiator.

An alternative wastage reaction is primary radical termination (see Chapter 6, E). Its importance increases as the kinetic chain length in the polymerization is reduced, either by decreasing [monomer] or by increasing [initiator]. It may be detected by kinetic methods and by end-group analysis in certain cases. It would cause the efficiency of initiation to fall as [monomer] is reduced; the resulting dependence of rate of initiation upon [monomer] could account for the order with respect to monomer for the overall process of polymerization being greater than unity in some cases. Direct studies of primary radical termination in the polymerization of styrene indicate that normally, however, it is not an important process, and that it could not be responsible for the efficiency of azo*iso*butyronitrile being only about 60%.

If the initiator gives highly stabilized radicals which are rather unreactive towards monomer, a significant proportion of them may be consumed by recombination in the body of the solution, or by primary radical termination even at quite high concentrations of monomer. Azobisdiphenylmethane is an initiator of this type and, as expected, its efficiency is low (Cohen and Wang, 1955).

A process which might be called *primary radical transfer* may be significant in connection with rates and efficiencies of initiation. Consider an initiator radical which can readily engage in a radical displacement reaction with one of the components of the reaction mixture

$$R\cdot + A.B \longrightarrow R.A + B\cdot \qquad (3)$$

to give a product radical which is rather unreactive towards monomer. An appreciable fraction of the B· radicals will be wasted, and the rate of initiation will decrease as [AB] is raised although [monomer] and [initiator] are held constant. Reaction of AB with a growing polymer radical would also cause an interruption to the growth of a polymer chain, and the substance would be classified as a retarder. There are examples, notably among the sulphur-containing initiators, of substances functioning both as initiators and retarders for radical polymerizations.

In some redox systems, radicals may be destroyed by reaction with a component of the redox couple, leading to inefficient initiation of polymerization.

C. Determination of Rates of Initiation

When sensitizers are used in polymerizations, rates of initiation are often deduced from measurements of the rate of decomposition of the sensitizers in a supposedly inert medium. The procedure is very uncertain because of doubts concerning the efficiencies of initiation, as explained in the previous section. The more reliable methods for determining rates of initiation can be divided into two groups:

(*a*) those in which the rate of formation of reactive centres is measured while the polymerization is in progress;
(*b*) those in which measurements on recovered polymer are used to determine the kinetic chain length during the polymerization.

One method of the first group involves the use of inhibitors. The length of the inhibition period is governed by [inhibitor] and the rate of generation of radicals. If each molecule of inhibitor is equivalent to one radical, the rate of initiation can be calculated from the relationship

$$\text{length of inhibition period} = \frac{\text{initial concentration of inhibitor}}{\text{rate of initiation}}.$$

It is assumed that both the inhibitor and monomer can suppress recombination of initiator radicals in the body of the solution, but that neither can interfere with geminate recombination.

One of the difficulties associated with the inhibitor method is in defining the exact length of the inhibition period. When [inhibitor] has been reduced to a very low value, the monomer and inhibitor can compete for radicals, and the rate of consumption of monomer therefore gradually builds up to its full value; kinetic analysis (see Chapter 7, A) indicates that the true inhibition period should be taken as the time required for the rate of polymerization to build up to 64·8% of its final value. Another uncertainty arises from the fact that reaction with an initiator radical may not convert the inhibitor into an unreactive substance but instead into one which can act as a retarder; a product of this type can contribute to the inhibition period, and so effectively each molecule of inhibitor is equivalent to more than one radical. This is probably the chief reason why it is usually found that if several inhibitors of different chemical types are used in a particular system, there are quite serious differences between the derived rates of initiation.

A particular advantage of the inhibitor method is that it can be used for both sensitized and unsensitized polymerizations: diphenylpicrylhydrazyl, for example, has often been used to measure the rate of production of radicals by high-energy radiations and by mechanical means. In photo reactions, difficulties may arise from the inhibitor acting as an internal filter.

Inhibitors may also be used to measure the rates of decomposition of initiators in inert solvents; several examples involving the use of diphenylpicrylhydrazyl were given in Chapter 2. These measurements give the rate of production of radicals available for reaction and not the total rate of decomposition; they are useful, therefore, for calculating rates of initiation if the inhibitors and monomers are equally effective in preventing the interaction of initiator radicals.

Retarders can be used in measurements of rates of initiation. If sufficient retarder is added to a polymerizing system to suppress mutual termination of reaction chains, it should be possible to relate the rate of initiation with the rate of consumption of retarder. The method depends upon some knowledge of the mechanism of the retardation, in particular the number of reaction chains stopped by one molecule of retarder; it is also necessary to have available a sensitive method for following either the fall in [retarder] during the reaction, or the rise in the concentration of a product formed from it. The retarder may be involved in primary radical termination or transfer, but the radicals concerned in these reactions would have initiated polymerization chains if the retarder had been absent and so no error arises.

A method depending upon the reaction of ferric chloride with a polymer radical

$$P\cdot + FeCl_3 \longrightarrow P.Cl + FeCl_2 \qquad (4)$$

is suitable in many cases (Bamford, Jenkins and Johnston, 1957). If in the steady state, all polymer radicals disappear by (4), the rate of appearance of ferrous ions in the system is equal to the rate of initiation in the absence of the ferric salt. The method can be applied to both sensitized and unsensitized polymerizations, but it can be used only in systems in which ferric chloride, or similar substance, can be dissolved.

In measurements of rates of initiation by methods of group (*b*), use is made of the relationship

$$\text{kinetic chain length } (\nu) = \frac{\text{rate of polymerization}}{\text{rate of initiation}}.$$

The kinetic chain length in a polymerization must be related to the number average molecular weight of the resulting polymer, and so measurements of overall rates of polymerization and molecular weights of polymers might lead to rates of initiation. The relationship between kinetic and molecular chain lengths is, however, not a universal one and depends upon the mechanism of the termination process and upon the frequency of transfer reactions. Uncertainty concerning this relationship is the main difficulty in the molecular-weight method for measuring rates of initiation.

If combination is the sole mode of termination and transfer can be neglected

$$\nu = \tfrac{1}{2}(\text{average molecular chain length})$$

since pairs of radicals unite to form a single molecule. In systems in which disproportionation occurs exclusively, or in which growing radicals disappear singly because of degradative transfer or certain other types of retardation,

$$\nu = \text{average molecular chain length.}$$

If linear and quadratic termination occur simultaneously, the kinetic chain length lies between these limiting values. Ordinary transfer reactions do not affect the rate of polymerization or the kinetic chain length, but they do reduce the average molecular weight of the resulting polymer and disturb the relationship between the kinetic and molecular chain lengths.

For determination of the molecular weight of the polymer, it is necessary to use a method which gives the number-average. It is essential that the molecules of comparatively small size in the hetero-disperse material make their proper contributions and that they are not lost during recovery and purification of the polymer. The magnitude of such losses must depend upon the natures of the polymer and the precipitant since these govern the extent to which low polymers are soluble. Calculations by Scanlan have been quoted by Ayrey and Moore (1959); for polymethyl methacrylate recovered by precipitation in methanol, about 2% of the polymer molecules are lost if the number-average molecular weight of the recovered polymer is about 5×10^5, but losses are larger if the average molecular weight is lower.

If a co-polymerization is being examined, the composition of the resulting co-polymer must be known, so that the average molecular weight determined experimentally can be converted into the number of monomer units in the average polymer molecule.

In a sensitized polymerization, the kinetic chain length is equal to the number of monomer units for each initiator fragment combined in the polymer, provided that

(i) initiation is solely by attachment of the initiator radical to a molecule of monomer, and
(ii) initiator does not enter the polymer by any other process.

Analysis for initiator combined in the polymer might, therefore, lead to a value for the kinetic chain length and then to the rate of initiation. The mechanism of the termination reaction is of no significance in this connection since it does not affect the ratio of the numbers of monomer units and initiator fragments in the polymer. Similarly, ordinary transfer reactions do not influence the result; if transfer is accompanied by retardation, the reduction in kinetic chain length compensates for the reduction in the rate of polymerization, and the rate of initiation is unaffected.

The only transfer reaction which can interfere is that involving initiator; it will be shown that allowance can quite easily be made for this effect, and that the process can be studied quantitatively.

Initiator fragments which may have entered the polymer as a result of primary radical termination cannot be distinguished from those entering as a result of initiation. The initiator fragment method may, therefore, give high values for the rate of initiation in some cases; if, however, primary radical transfer occurs, low results are obtained because an initiator fragment does not become incorporated in polymer. The effect of primary radical transfer can be observed even if the product radical initiates with 100% efficiency, because the initiator fragment method gives the rate of initiation by a particular species of radical.

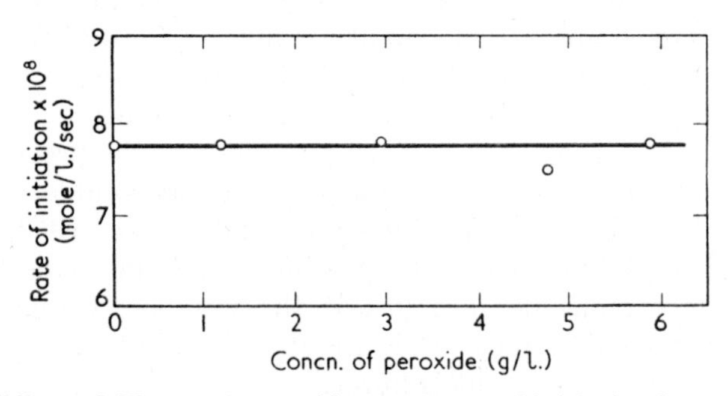

FIG. 3.3. Effect of dibenzoyl peroxide upon rate of initiation by azo*iso*butyronitrile at 1·00 g/l. in polymerization of styrene at 60°C.

The initiator-fragment method for determining rates of initiation obviously is applicable to sensitized polymerizations only, and the initiator must be such that accurate analyses for very low concentrations of end-groups in polymers are possible. If the fragment contains functional groups, conventional analytical methods of very high sensitivity may be used but tracer techniques are the most generally applicable (Bevington, 1958).

As in the molecular-weight method, errors may arise from loss of polymer molecules during recovery and purification of the polymer. The end-groups must be firmly bound to the polymer molecules so that they do not become detached during the isolation of the polymer. The remarks concerning rates of initiation in co-polymerizations, already made in connection with the molecular weight method, apply here also. When using a labelled initiator, the position of the labelling atom in the molecule must be chosen with care for two reasons. The first of these is that the radical formed initially may subsequently dissociate into a smaller radical and a stable molecule. The second reason is that there may be an isotope effect in the dissociation of the initiator; generally these effects are small, but it is advisable to label the initiator at a position such that a labelled atom is not directly attached to the valency bond which is broken.

The composition, but not the structure, of the initiating radical must be known in order to interpret the analyses; for example, when using azo*iso*butyronitrile it is immaterial whether the initiating radical is $(CH_3)_2C(CN)\cdot$ or $(CH_3)_2C{:}C{:}N\cdot$ since the two forms have identical compositions. With this proviso, it should be emphasized that a special advantage of the initiator-fragment method is that it is possible to study

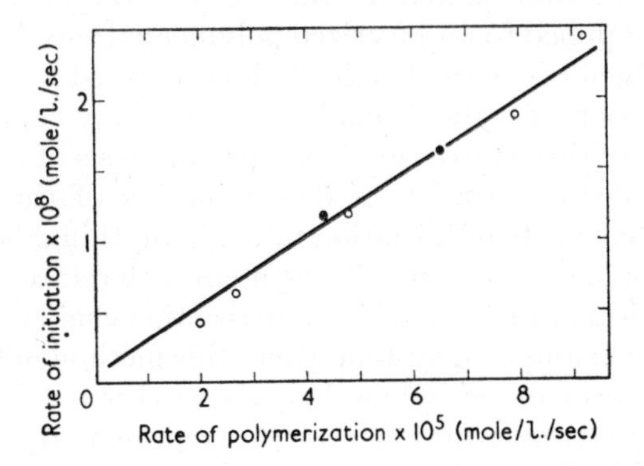

FIG. 3.4. Rate of initiation by bis(4-methoxy-3,5-dibromobenzoyl) peroxide at 2·93 g/l. in polymerization of styrene at 60°C in presence of other initiators. ○ azo*iso*butyronitrile; ● dibenzoyl peroxide.

initiation by one particular type of radical when, in the actual polymerization, there is more than one initiation process. It is possible, therefore, to study initiators in systems where direct thermal or photo-polymerization is significant.

This advantage can be appreciated by considering polymerizations involving mixtures of initiators. The addition of a peroxide to a system containing styrene and azo*iso*butyronitrile at fixed concentrations does not affect the rate of initiation by the azo compound (see Fig. 3.3) (Bevington and Lewis, 1960*a*). The second initiator does not influence either the rate at which the azo compound dissociates to radicals, or the efficiency with which monomer captures these radicals. It is possible that with very high concentrations of the second initiator, primary radical termination might become important, but this would not be detected in this method of working. If the labelled initiator, however, is one susceptible to transfer, a different result is obtained (see Fig. 3.4). As the concentration of the second initiator is raised, the stationary concentration of polymer radicals increases, and transfer reactions involving the labelled initiator become more frequent. The true rate of initiation by the labelled initiator is the intercept on the R_i axis, since this corresponds to a system in which [growing radicals] is zero, and transfer to initiator is, therefore, completely eliminated; the procedure is considered again in Chapter 5, E. The method should be applicable to initiators of other kinds, for example some of the sulphur-containing initiators, which may act as transfer agents and retarders, and also to those such as benzoin and derivatives which can act as co-monomers in photo-polymerizations (see Section B.*1* of Chapter 4).

A method somewhat similar to the initiator-fragment method, but which might be applied to unsensitized polymerizations, involves the use of a retarder which becomes chemically incorporated in the polymer. If mutual termination of reaction chains is completely suppressed, and if the mechanism of the retardation is known, analysis of the polymer for combined retarder can lead to a determination of the kinetic chain length. The derived rate of initiation must, if anything, be less than the rate of initiation in the corresponding system with retarder absent, since some of the radicals which might initiate reaction chains may react with the retarder. An example of system where this method might be applied is the polymerization of methyl methacrylate in the presence of *p*.benzoquinone; the relevant reactions can be represented as (Bevington, Ghanem and Melville, 1955*a*)

$$\mathrm{P\cdot + O{=}C_6H_4{=}O \longrightarrow P{-}O{-}C_6H_4{-}O\cdot} \qquad (5)$$

$$\mathrm{P{-}O{-}C_6H_4{-}O\cdot + P\cdot \longrightarrow P{-}O{-}C_6H_4{-}O{-}P} \qquad (6)$$

and

$$\nu = \frac{1}{2}\cdot\frac{\text{no. of monomer units in polymer}}{\text{no. of quinone molecules in polymer}}.$$

This, and other cases, are considered in Chapter 7. Generally there are considerable differences between the reactivities of various polymer radicals towards a particular retarder, so that each system must be considered separately.

A comparative method for measuring rates of initiation is based on the fact that for a given set of conditions, a particular rate of polymerization must correspond to a certain rate of initiation independent of the nature of the initiation process. It has been used, for example, to determine the rate of decomposition of ethylene glycol dinitrate into radicals at fairly low temperatures (Hicks, 1956). The method is valid only if none of the initiators involved in the comparison engage in degradative transfer. In some cases, the rate of initiation is calculated from the observed rate of polymerization using accepted values for the velocity constants for chain growth and termination; the original determinations of these velocity constants must, however, have involved measurements of rate of initiation.

D. Relative Reactivities of Monomers towards Primary Radicals

1. Unstable Primary Radicals

The reactivities of monomers towards certain initiator radicals can be compared by competitive studies of various types. One such method is applicable to radicals which may themselves dissociate; monomers can interfere in these decompositions to extents depending upon their concentrations and their reactivities towards the unstable radicals. The primary radical must be one which, at the working temperature, survives on the average for at least 10^{-9} sec (see Section B) in an inert solvent, so that the numbers of radicals decomposing and being captured by monomer are comparable. The technique is particularly suitable for the benzoyloxy radical, and this case will be discussed in some detail to illustrate the general principles. The velocity constants for reactions of monomers towards the unstable radical are compared by reference to the velocity constant for the unimolecular decomposition of the radical.

When dibenzoyl peroxide is used an an initiator, the reactions

$$C_6H_5.CO.O\cdot + M \longrightarrow C_6H_5.CO.O.M\cdot \qquad (7)$$

$$C_6H_5\cdot + M \longrightarrow C_6H_5.M\cdot \qquad (8)$$

occur. Changing [monomer] alters the relative importances of these reactions but, over a wide range, the total rate of initiation is constant (Bevington, 1957). Evidently the benzoyloxy radicals which escape (7) decompose thus:

$$C_6H_5.CO.O\cdot \longrightarrow C_6H_5\cdot + CO_2 \qquad (9)$$

and the phenyl radicals so formed are captured by monomer.

The fraction (x) of benzoyloxy radicals captured by monomer is given by

$$x = \frac{\text{rate of (7)}}{\text{rate of (7)} + \text{rate of (9)}} = \frac{k_7[B\cdot][M]}{k_7[B\cdot][M] + k_9[B\cdot]}$$

where [B·] = stationary concentration of benzoyloxy radicals,
[M] = concentration of monomer,

so that
$$\frac{1}{x} = 1 + \frac{k_9}{k_7[M]}$$

C*

If all the phenyl radicals are captured by monomer, in the steady state the rates of (8) and (9) are equal, and

$$x = \frac{\text{rate of (7)}}{\text{rate of (7)} + \text{rate of (8)}}$$

$$= \frac{\text{the no. of benzoyloxy end-groups in polymer}}{\text{(no. of benzoyloxy end-groups)} + \text{(no. of phenyl end-groups)}}.$$

The fraction x can be determined in two ways, both involving labelled peroxide and measurements on the recovered polymers. The first requires the use of two types of labelled peroxide, the one labelled at the carboxyl carbon atoms only and the other labelled in the benzene rings only; it is possible to determine the rates of initiation by benzoyloxy and phenyl

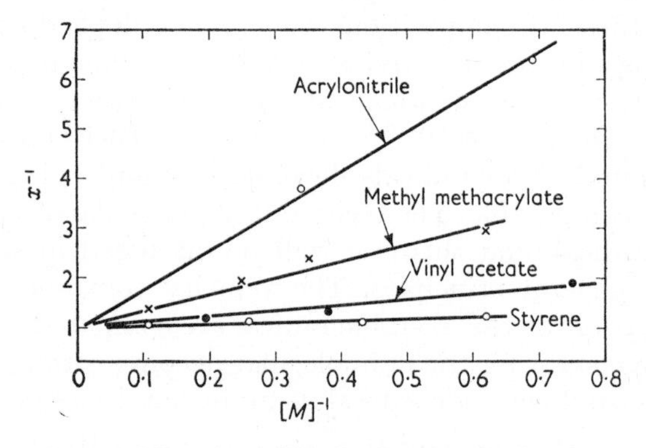

FIG. 3.5. Dependence of fraction of benzoyloxy radicals captured by monomer upon [monomer] at 60°C.

radicals separately (Bevington, 1957). In the other method of working, it is necessary to use only one type of labelled peroxide; this may be labelled uniformly or in the benzene rings only. The benzoyloxy end-groups, being attached to the polymer chains by ester linkages, can be hydrolysed off the polymers without difficulty. Comparison of the specific activities of a polymer before and after hydrolysis leads to a value for the relative numbers of the two types of end-group (Bevington and Brooks, 1956). For monomers containing groups sensitive to hydrolysis, it may be necessary to make allowance for changes in composition of the polymer resulting from reactions of the monomer units.

Plots of $1/x$ against $[\text{monomer}]^{-1}$ for some monomers are shown in Fig. 3.5; derived values of k_9/k_7 at 60°C are displayed in Table 3.1. The relative values of k_7 are significant in connection with the role of polar effects in radical-addition processes (see Section B.*4* of Chapter 4).

According to the treatment developed here, x should be independent of [initiator] if [monomer] and the temperature are fixed. In some systems, x increases with rising concentration of initiator; this is due either to transfer to initiator or primary radical termination. These effects are usually very small; if necessary, they can be allowed for by extrapolating all results to zero concentration of initiator.

The uncertainty about the results for acrylonitrile arises from the fact that with this monomer the initiator-fragment method gives rather low rates of initiation, probably because of primary radical transfer (Barson, Bevington and Eaves, 1958). If the benzoyloxy radical is involved in this process, the values of x and k_9/k_7 are unaffected; if, however, the phenyl radical engages in this type of transfer, the expression for x must be

TABLE 3.1

REACTIVITIES OF MONOMERS TOWARDS THE BENZOYLOXY RADICAL AT 60°C

Monomer	Diluent	k_9/k_7 (mole/l.)	Relative value of k_7
2,5-dimethyl styrene	benzene	0·2	2·0
styrene	,,	0·4	1·0 (standard)
2,4,6-trimethyl styrene	,,	0·6	0·67
vinyl acetate	,,	1·1	0·36
methyl methacrylate	,,	3·3	0·12
acrylonitrile	dimethylformamide	$\geqslant$8·0	$\leqslant$0·05

modified since the rate of incorporation of these radicals in polymer is less than the rate at which they are formed.

Values of k_9/k_7 have been determined at 80°C also for certain monomers. For styrene, 2,5-dimethyl styrene and 2,4,6-trimethyl styrene, the values of (E_9-E_7) are respectively 6·6, 4·7 and 3·4 kcal/mole, so that the energy of activation for the dissociation of the benzoyloxy radical is probably in the region of 12 kcal/mole (Bevington and Toole, 1958).

When dibenzoyl peroxide is used as a photosensitizer, plots of $1/x$ against $[\text{monomer}]^{-1}$ are linear but do not pass through the point $(x^{-1}=1;\ [\text{M}]^{-1}=0)$ (see Fig. 3.6). The primary step for the photodissociation ought to be written as

$$(\text{C}_6\text{H}_5.\text{CO}.\text{O})_2 \longrightarrow 2(1-f)\text{C}_6\text{H}_5.\text{CO}.\text{O}\cdot + 2f\text{C}_6\text{H}_5\cdot + 2f\text{CO}_2 \qquad (10)$$

where f is less than unity. Simple kinetic analysis shows that in this case

$$\frac{1}{x} = \frac{1}{1-f} + \frac{k_9}{k_7[\text{M}](1-f)} \qquad (11)$$

From Fig. 3.6, f is about 0·3, and it is apparently independent of temperature. Values for k_9/k_7 obtained in photochemical and thermal reactions are in reasonable agreement, indicating that the benzoyloxy radicals generated during the photolysis of dibenzoyl peroxide in the near ultra-violet are not in an excited state and behave in the way they would if generated thermally.

Comparisons of the reactivity and stability of the benzoyloxy radical generated thermally could be made over a wider range of temperatures by using other sources for the radical. Thus the decomposition of N-nitrosobenzanilide or the amine-induced decomposition of dibenzoyl peroxide might be used at comparatively low temperatures; the thermal

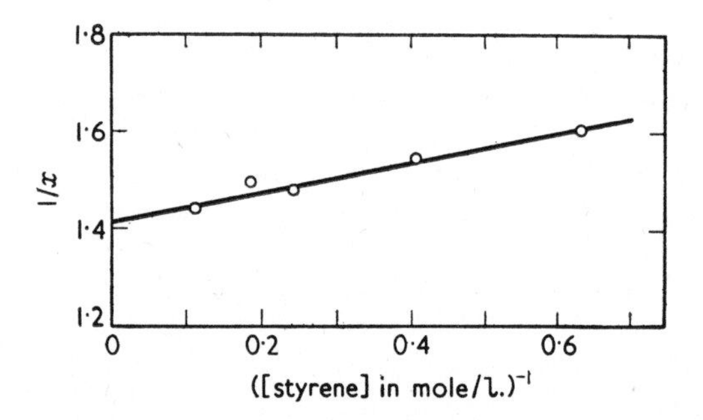

Fig. 3.6. Effect of [styrene] upon fraction of benzoyloxy radicals captured by monomer at 25°C—radicals generated by photolysis of peroxide.

dissociation of *tert.*butyl perbenzoate might be used at temperatures rather too high for the convenient use of dibenzoyl peroxide.

Similar studies have been made of the reactivities of monomers towards substituted aroyloxy radicals. Introduction of substituents into the benzene ring can cause profound changes in the value of k_9/k_7 (see Table 3.2), but the arrangement of monomers in order of reactivity towards the *p*.methoxybenzoyloxy radical, for example, is the same as that when the benzoyloxy radical is used as the reference radical (Allen and Bevington, 1960*a*).

Competitive studies of this type can be extended to mixtures of monomers which give co-polymers; in this case, the alternative reactions for the benzoyloxy radical, for example, are

$$C_6H_5.CO.O\cdot \longrightarrow C_6H_5\cdot + CO_2 \quad (9)$$

$$C_6H_5.CO.O\cdot + M_a \longrightarrow C_6H_5.CO.O.M_a\cdot \quad (12)$$

$$C_6H_5.CO.O\cdot + M_b \longrightarrow C_6H_5.CO.O.M_b\cdot \quad (13)$$

Of the benzoyloxy radicals, the fraction incorporated in the co-polymer is given by

$$\frac{x}{1-x} = \frac{k_{12}[\mathrm{M}_a]}{k_9} + \frac{k_{13}[\mathrm{M}_b]}{k_9} \tag{14}$$

This equation has been found valid for mixtures of styrene and methyl methacrylate (Allen and Bevington, 1960*b*). Its chief value is for determining the relative reactivities towards the reference radical of unsaturated substances, such as stilbene, which do not form homo-polymers but which enter in co-polymerizations; such substances react quite readily

TABLE 3.2

COMPARISONS OF THE STABILITIES OF AROYLOXY RADICALS AND THEIR REACTIVITIES TOWARDS STYRENE

Radical	k_9/k_7 in mole/l. at 60°C	k_9/k_7 in mole/l. at 80°C
benzoyloxy	0·4 (*a*)	0·7 (*b*)
p.methoxybenzoyloxy	0·02 (*c*)	0·05 (*c*)
m.methoxybenzoyloxy	0·3 (*d*)	0·6 (*d*)
m.bromobenzoyloxy	$\geqslant$ 0·3 (*e*)	$\geqslant$ 0·5 (*e*)
3,5-dibromo-4-methoxybenzoyloxy	$\geqslant$ 0·06 (*e*)	$\geqslant$ 0·4 (*e*)

(*a*) Bevington (1957)
(*b*) Bevington and Toole (1958)
(*c*) Bevington, Toole and Trossarelli (1958*a*)
(*d*) Bevington, Toole and Trossarelli (1958*b*)
(*e*) Bevington, Toole and Trossarelli (1959)

with primary radicals, but the velocity constant for growth of a homo-polymer is vanishingly small.

Initiators of several distinct types dissociate to radicals which may subsequently decompose, and competitive methods might be used for comparing the reactivities of monomers towards the radicals first formed. It would be necessary to use initiators isotopically labelled at suitable positions. Some of the initiators which might be used in this way are listed in Table 3.3, but it appears that only di-*tert.*butyl peroxide has been examined in any detail (Allen and Bevington, 1960*a*).

In studies of this type, it is necessary to perform experiments over a range of concentrations of scavenger; it is usually assumed that the solvent plays no part beyond acting as a diluent. There is evidence however (Russell, 1959) that the *tert.*butoxy radical may form π-complexes with aromatic hydrocarbons, and that in such complexes the radical may exhibit reduced but more selective reactivity. For the reactions

$$(CH_3)_3C.O\cdot \longrightarrow (CH_3)_2CO + CH_3\cdot \quad (15)$$

$$(CH_3)_3C.O\cdot + \text{cyclo-}C_6H_{12} \longrightarrow (CH_3)_3C.OH + \text{cyclo-}C_6H_{11}\cdot \quad (16)$$

$$(CH_3)_3C.O\cdot + CH_2{:}CH.C_6H_5 \longrightarrow (CH_3)_3C.O.CH_2.CH(C_6H_5)\cdot \quad (17)$$

the values of k_{15}/k_{16} and k_{15}/k_{17} depend upon the nature of the diluent. Clearly this possibility must be considered in all competitive studies. In the case of the benzoyloxy radical with styrene, there were no changes resulting from the replacement of benzene by dimethylformamide as diluent, indicating that complex formation may be ignored in this case.

TABLE 3.3

EXAMPLES OF INITIATORS GIVING UNSTABLE RADICALS

Initiator	Radical decomposition
di-*tert.*butyl peroxide	$(CH_3)_3C.O\cdot \longrightarrow (CH_3)_2CO + CH_3\cdot$
di-*iso*propyl peroxydicarbonate	$(CH_3)_2CH.O.CO.O\cdot \longrightarrow (CH_3)_2CH.O\cdot + CO_2$ $(CH_3)_2CH.O\cdot \longrightarrow CH_3.CHO + CH_3\cdot$
di-alkyl azodicarboxylate	$R.O.C.O\cdot \longrightarrow R\cdot + CO_2$ or $\longrightarrow R.O\cdot + CO$
tetra-alkyl thiuram disulphide	$R_2N.CS.S\cdot \longrightarrow R_2N\cdot + CS_2$

When using labelled radicals in competitive studies, it is necessary to be aware of the possibility of isotope effects is dissociations such as (15). During the dissociation of the radical $^{14}CH_3.C(CH_3)_2.O\cdot$, at about 130°C, the chance of a $^{14}C—^{12}C$ bond breaking is about 80% of the chance of a $^{12}C—^{12}C$ bond dissociating and there is therefore a tendency for the heavy carbon atom to appear in the acetone rather than in the methyl radical; further, the overall velocity constant for dissociation of the radicals containing carbon-14 is only about 90% of the corresponding quantity for unlabelled radicals. These effects can have small but significant influences on ratios such as k_{15}/k_{17} (Allen and Bevington, 1960*a*).

2. Competition Between Scavengers

If radicals are generated in mixtures of substances with which they may react, there are competitions between the various possible reactions.

If one of the reactive substances is a monomer and the other a transfer agent, the reaction scheme is

$$R\cdot + M \longrightarrow R.M\cdot \tag{18}$$

$$R\cdot + S.H \longrightarrow R.H + S\cdot \tag{19}$$

If the proportions of the radicals reacting in these alternative ways can be determined, by product analysis, it is possible to express k_{18} for various monomers in terms of k_{19}.

The procedure has been used extensively (Szwarc, 1955) in connection with methyl radicals produced from di-acetyl peroxide, using *iso*-octane as the reference compound S.H. The gaseous products of the decomposition of di-acetyl peroxide in *iso*-octane are methane, ethane and carbon dioxide; the ethane is formed almost entirely by geminate recombination of methyl radicals and the methane by (19). If an unsaturated compound is present in the reaction mixture, the yields of carbon dioxide and ethane are unaffected, showing that the total rate of production of methyl radicals and the rate of their geminate recombination are unaltered; the yield of methane is depressed, however, because (18) competes with (19). If all the methyl radicals are consumed in either (18) or (19), the fraction (y) of the "available" methyl radicals reacting with the octane is given by

$$y = \frac{k_{19}[CH_3\cdot][SH]}{k_{19}[CH_3\cdot][SH]+k_{18}[CH_3\cdot][M]} \tag{20}$$

where $[CH_3\cdot]$ = the stationary concentration of methyl radicals, and [SH] and [M] = the concentrations of *iso*-octane and monomer respectively.

The fraction y is determined experimentally as

$$\frac{\text{no. of moles of } CH_4 \text{ produced}}{(\text{no. of moles of } CO_2 \text{ produced}) - 2(\text{no. of moles of } C_2H_6 \text{ produced})}$$

since a molecule of CO_2 accompanies each methyl radical, but some of these radicals undergo geminate recombination to give ethane and so are not available for the competition between (18) and (19). If average values are assigned to [SH] and [M], k_{18}/k_{19} can be calculated. Some of the results of studies of this type are shown in Table 3.4.

Usually monomers are considerably more reactive than *iso*-octane towards the methyl radical, and y can be determined accurately only at tow concentrations of the monomer; the resulting polymer chains, therefore, are very short. A significant proportion of the methyl radicals may react with polymer radicals or with *iso*-octane radicals, so disturbing the relationship (20).

The procedure outlined above has been applied also to study of ethyl radicals, generated from di-*n*.propionyl peroxide, and *n*.propyl radicals, derived from di-n.butyryl peroxide (Smid and Szwarc, 1956). The patterns of behaviour of these radicals are very similar to that for the methyl radical.

In cases where [monomer] is fairly high, the competition between primary radical transfer (Chapter 5, F) and initiation might be studied by analysis of the resulting polymer for initiator fragments. If the total rate of production of "available" R· radicals is unaffected by the presence

TABLE 3.4

RELATIVE REACTIVITIES OF MONOMERS TOWARDS THE METHYL RADICAL

Monomer	k_{18}/k_{19} at 65°C	k_{18}/k_{19} at 85°C
acrylonitrile	1540	—
methyl methacrylate	1420	960
styrene	792	651
vinyl acetate	37	28
ethylene	34·1	35·5
allyl acetate	8·4	5·7

of the transfer agent, and if all these radicals are consumed either in (18) or (19),

$$\frac{\text{rate of initiation by R· radicals}}{\text{rate of production of R· radicals}} = \frac{R_i}{R_{i_0}} = \frac{k_{18}[\text{R·}][\text{M}]}{k_{18}[\text{R·}][\text{M}]+k_{19}[\text{R·}][\text{SH}]}$$

so that

$$\frac{R_{i_0}}{R_i} = 1+\frac{k_{19}[\text{SH}]}{k_{18}[\text{M}]}$$

where R_{i_0} = rate of initiation in absence of transfer agent. Determination of the rate of initiation, R_i, in the presence of the transfer agent should then allow k_{19}/k_{18} to be determined. Using a series of monomers with a particular transfer agent could lead to a set of relative values for the velocity constant k_{18}; experiments with various transfer agents and a given (monomer–initiator) pair could lead to a set of relative values for k_{19}, the reaction between the initiator radical and the chosen monomer being the reference standard.

In practice, this method depending upon measurements of rates of initiation only is of very limited application. The transfer agent must be fairly reactive in order to give values of R_{i_0}/R_i appreciably greater than 1 at moderate concentrations of the additive; at the same time, transfer to growing polymer radicals must not be very frequent, otherwise the molecular weight of the resulting polymer is very low and serious errors may result from losses during recovery and purification. The method can be successful only if k_{19}/k_{18} for the reactions of the primary radical is appreciably greater than the corresponding ratio for the reactions involving the growing polymer radical. A system in which primary radical transfer can interfere with the normal initiation process is that of styrene with azo*iso*butyronitrile and carbon tetrabromide; $(CH_3)_2C(CN)Br$ can be detected in the reaction mixture as the product of the primary radical transfer.

Primary radical transfer may be significant when persulphates are used as initiators in aqueous systems; there is competition between

$$SO_4\cdot^- + H_2O \longrightarrow HSO_4^- + \cdot OH \tag{21}$$

and

$$SO_4\cdot^- + M \longrightarrow {}^-SO_4.M\cdot \tag{22}$$

If the system is homogeneous and any hydroxyl radicals formed in (21) are subsequently captured by monomer:

$$\frac{\text{no. of OH end-groups in polymer}}{\text{no. of } SO_4^- \text{ end-groups}} = \frac{k_{21}[SO_4\cdot^-][H_2O]}{k_{22}[SO_4\cdot^-][M]} = \frac{k_{21}[H_2O]}{k_{22}[M]}$$

Determination of the relative numbers of the two types of end-group would therefore give a value for k_{21}/k_{22}, provided that OH end-groups do not also enter the polymer by transfer reactions involving polymer radicals and water. If this condition is satisfied, experiments with a series of monomers in aqueous solution would lead to a comparison of the reactivities of the monomers towards the sulphate radical-ion using (21) as the reference reaction.

This treatment could not be applied to a system in which the monomer is distributed non-uniformly, say as droplets suspended in the water. Then, the number of radical-ions engaging in (21) depends upon the average separation between the droplets, since this governs the average interval between the generation of a radical-ion in the aqueous phase and the entry of a radical into a droplet of monomer.

Persulphate labelled with sulphur-35 has been used on a number of occasions to initiate polymerizations in emulsions, and it has been shown that the resulting polymers contain activity. In some cases it is likely that not all occluded active material was eliminated from the polymer, so that the results must be suspect. Palit (1960) has been unable to detect ionic end-groups in polymethyl methacrylate prepared in emulsion using persulphate as initiator. This could be explained if (21) is very pronounced, and evidently these systems require careful re-examination.

Another competitive method may be applicable when redox systems are used to generate radicals; there may be competition between the monomer and the reducing ion for reaction with the radical. A typical reaction scheme (see Section B of Chapter 2) is:

$$R.O.O.H+Fe^{++} \longrightarrow Fe^{+++}+OH^{-}+R.O\cdot \quad (23)$$

$$R.O\cdot+Fe^{++} \longrightarrow Fe^{+++}+RO^{-} \quad (24)$$

$$R.O\cdot+M \longrightarrow R.O.M\cdot \quad (25)$$

If these are the only reactions which need to be considered, the ratio

$$\frac{\text{no. of moles of } Fe^{++} \text{ oxidized}}{\text{no. of moles of hydroperoxide consumed}}$$

should change from 2 to 1 as [monomer] is increased from a very low to a very high value. This dependence of the stoichiometric ratio upon [monomer] forms the basis for comparison of the reactivities of monomers towards certain initiating radicals.

The method, introduced by Baxendale, Evans and Park (1946) for study of the $\cdot OH$ radical produced from hydrogen peroxide, was developed by Orr and Williams (1955) and used with substituted cumyloxy radicals and the sulphate radical-ion. It was applied to the $\cdot NH_2$ radical generated from hydroxylamine (Davis, Evans and Higginson, 1951). The method is restricted to monomers appreciably soluble in water.

In the simplest treatment, it is supposed that the only reactions of the hydroperoxide, the alkoxy radical and the ferrous ion are (23), (24) and (25). Other reactions of possible significance are:

(*a*) reaction between the hydroperoxide and R.O· radicals:

$$R.O\cdot+R.O.O.H \longrightarrow R.OH+R.O.O\cdot$$

(*b*) reaction between the hydroperoxide and polymer radicals, i.e. transfer to initiator;

(*c*) capture of R.O· radicals by polymer radicals, i.e. primary radical termination;

(*d*) conversion of the R.O· radicals into radicals of other types by decomposition or by primary radical transfer, e.g.

$$C_6H_5.C(CH_3)_2.O\cdot \longrightarrow C_6H_5.CO.CH_3+CH_3\cdot$$

$$SO_4^{-}\cdot+H_2O \longrightarrow HSO_4^{-}+\cdot OH$$

(*e*) reaction between polymer radicals and ferrous ions

$$P\cdot+Fe^{++}+H_2O \longrightarrow P.H+Fe^{+++}+OH^{-}$$

Orr and Williams made allowance for (*a*) and (*e*) and obtained results shown in Tables 3.5 and 3.6.

The pattern of results in Table 3.6 is indistinct; from Table 3.5, however, it is evident that both $(E_{25}-E_{24})$ and A_{25}/A_{24} decrease as the

TABLE 3.5

DATA ON REACTIONS OF SOME SUBSTITUTED CUMYLOXY RADICALS

Para substituent in radical	Monomer	A_{25}/A_{24}	$(E_{25}-E_{24})$ kcal/mole
hydrogen	acrylonitrile	2×10^{-3}	0
	methyl methacrylate	$0\cdot9\times10^{-3}$	3
	methyl acrylate	1×10^{-3}	8·6
tert.butyl	acrylonitrile	2×10^{10}	8
	methyl methacrylate	1×10^{4}	—
nitro	acrylonitrile	2×10^{-4}	−2·5
	methyl methacrylate	4×10^{-4}	−1·9
	methyl acrylate	4×10^{-5}	−3·4

para substituent in the cumyloxy radical becomes more electronegative; there is a relationship between these quantities and the Hammett substituent constant. A value for k_{25}/k_{24} for reactions involving one of these monomers and another *p*.substituted cumyloxy radical could therefore be predicted.

TABLE 3.6

RELATIVE REACTIVITIES OF MONOMERS TOWARDS CUMYLOXY RADICALS

Radical	Temp. (°C)	Relative value of k_{25}: methyl methacrylate	acrylo-nitrile	methyl acrylate
cumyloxy	25	1	0·65	0·35
	48	1	2·00	1·33
	59	1	0·57	0·86
p.nitro-cumyloxy	40	1	1·25	1·00
	48	1	1·25	1·12
	59	1	1·33	1·05
p.*tert*.butyl-cumyloxy	0	1	5·62	—
	14	1	2·70	2·23

For the sulphate radical-ion, the reaction corresponding to (24) involves two ions. The values for k_{25}/k_{24}, therefore, depend upon the ionic strength; the results in Table 3.7 have been corrected to zero ionic strength by the Bronsted-Bjerrum equation.

[Monomer] is normally considerably greater than [ferrous ion], but the values of k_{25}/k_{24} for these monomers are so small that the rates of (24) and (25) may be comparable, leading to a significant loss of radicals from the system. If the values of k_{25}/k_{24} are reliable, it means that the redox combination of persulphate and the ferrous ion may be inefficient for the initiation of the polymerizations of these monomers. Wastage of radicals by (24) is reduced by adding to the system a complexing agent so that the concentration of free ferrous ions is perhaps only 10^{-9} M.

Orr and Williams showed that the stoichiometric ratio for the reaction between the ferrous ion and cumene hydroperoxide or the persulphate

TABLE 3.7

COMPARISONS OF VELOCITY CONSTANTS FOR REACTIONS OF THE SULPHATE RADICAL-ION

Monomer	A_{25}/A_{24}	$(E_{25}-E_{24})$ kcal/mole	k_{25}/k_{24} at 25°C
acrylonitrile	1×10^{-16}	-17	$3{\cdot}9\times11^{-4}$
methyl methacrylate	6×10^{-22}	-26	$7{\cdot}7\times10^{-3}$
methyl acrylate	7×10^{-9}	-7	$1{\cdot}1\times10^{-3}$

ion is affected by the presence of methanol. The radical formed in the primary reaction must react with methanol to give another radical unreactive towards the ferrous ion; it is thought that

$$R.O\cdot + CH_3OH \longrightarrow R.OH + \cdot CH_2OH \qquad (26)$$

is another reaction which may compete with (25). In certain emulsion polymerizations, methanol may be added to the system to increase the solubility of the peroxide in the aqueous phase; in these cases, the *tert.*-butylcumene hydroperoxide is preferred because (26) is of little importance for the radical derived from this peroxide.

The procedure just discussed essentially involves a determination of the number of radicals which do not initiate polymerization but instead engage in (24); the number of radicals reacting according to (25) is found by difference. An alternative procedure would involve direct measurement of the rate of (25) by analysis of the resulting polymer for end-groups derived from the peroxide. It would be necessary to choose conditions to favour the production of polymers of fairly high molecular weight for analysis. In the simple case where only (23), (24) and (25) have to be considered, the fraction (x) of the R.O• radicals captured by monomer could be determined as

$$\frac{\text{rate of incorporation of R.O• radical in polymer}}{\text{rate of formation of these radicals by (23)}}.$$

The usual treatment for the competition between (24) and (25) shows that

$$\frac{1}{x} = 1 + \frac{k_{24}[\mathrm{Fe}^{++}]}{k_{25}[\mathrm{M}]} \qquad (27)$$

and k_{25}/k_{24} could be evaluated. There might be additional terms on the right-hand side of (27) corresponding to reactions of R.O• with other components of the system; each of these terms would be of the form $k[\text{component}]/k_{25}[\mathrm{M}]$, and could be evaluated from sets of experiments in which the concentration of each component is in turn varied. Reaction of polymer radicals with ferrous ions would not affect the results. It might be possible also to correct for transfer to peroxide, primary radical termination and the possibility of the primary radical being converted into another radical, for example by decomposition, using procedures discussed in connection with competitive studies involving the benzoyloxy radical.

CHAPTER 4

Growth Reactions

A. GENERAL FEATURES

1. Introduction

The growth reaction in a radical polymerization is represented by the general equation

$$P_n\cdot + M \longrightarrow P_{n+1}\cdot \tag{1}$$

and is of the same general type as the initiation reaction. In kinetic analyses of polymerizations, it is usually supposed that the value of k_p is independent of the size of the radical. In more general treatments (Gee and Melville, 1944; Bamford, Barb, Jenkins and Onyon, 1958), variation of the reactivity of a polymer radical with its size is permitted. The expressions which can be derived for the overall rate of polymerization and the molecular weight of the resulting polymer, involve ratios of the velocity constants for the elementary reactions; it is supposed that the individual velocity constants may alter but that the appropriate ratios remain independent of radical size. Some of the most recent reports on effects of radical size upon reactivity have been concerned with polymerizations in the presence of bromotrichloromethane which can act as powerful transfer agent and also as a photosensitizer (Robb and Vofsi, 1959; Bengough and Thomson, 1960). For small polymer radicals, the effect of size upon reactivity results from the influence of the group at the non-reacting end of the radical, and it is probable that this can be transmitted through three or four monomer units at the most. The magnitude of the effect upon the rate of reaction must be related to the nature of the molecule with which the radical reacts, and so the reactivities of the radicals in their various reactions may be affected to different extents; a dependence of ratios such as k_f/k_p and k_t/k_p upon radical size can therefore be expected, but it is likely to be significant only for very short polymer radicals.

In the simple case in which

(*a*) the rate of production of "available" radicals is independent of [monomer];
(*b*) all such radicals are captured by monomer;
(*c*) termination is solely by the interaction of polymer radicals;

the stationary concentration of polymer radicals is independent of [monomer], being governed by the relationship

$$R_i = k_t[\mathrm{P}\cdot]^2$$

The overall process of polymerization should, therefore, be first order with respect to monomer, since the rate of growth is $k_p[\mathrm{P}\cdot][\mathrm{M}]$, and it is only in this reaction that significant quantities of monomer are consumed. It is usual, for simplicity, to deal with concentrations of monomer rather than thermodynamic activities, although the latter have been determined and used to a limited extent (Walling, Briggs and Mayo, 1946).

Direct proportionality between rate of polymerization and [monomer] has been found for the sensitized polymerization of methyl methacrylate in benzene and various other diluents. For some monomers higher orders are found, for example with styrene in benzene the order is 1·2, and with vinyl acetate in benzene the order has a very high value at high concentrations of monomer, dropping to about two for dilute solutions. In these cases there is probably no abnormality in the growth reaction itself, but the stationary concentration of growing radicals depends upon [monomer].

Various explanations for this effect are considered in detail in other sections; there are two general approaches, viz.

(*a*) those based on the idea that the rate of initiation is not independent of [monomer];

(*b*) those in which it is supposed that termination is not solely by the interaction of polymer radicals.

Dependence of the rate of initiation and of the stationary concentration of growing centres upon [monomer] would be found if recombination of primary radicals could compete with capture of these radicals by monomer, or if an appreciable fraction of the primary radicals engaged in termination processes. In both cases, a reduction in [monomer] would favour the wastage of primary radicals so that the rate of initiation would decrease and the stationary concentration of growing radicals would fall; the overall order of the polymerization with respect to monomer would consequently be greater than 1. The explanations in group (*b*) suppose that the growing radicals engage in transfer reactions with the diluent to give comparatively unreactive radicals, some of which are removed by termination processes. This type of termination obviously becomes more pronounced as [diluent] is increased and [monomer] is reduced, causing the stationary concentration of polymer radicals to fall; the overall order with respect to monomer, therefore, is greater than unity.

In direct thermal or photo-polymerizations, high orders with respect to monomer are easily explained since the rate of production of radicals must be a function of [monomer]. Conflicting results in some of these cases can probably be attributed to failure by some workers to purify the reactants completely.

Values of k_p at 60°C selected by Bamford, Barb, Jenkins and Onyon (1958) for some of the commoner monomers are shown in Table 4.1. The precise numerical values may be open to doubt but differences between the results of various groups of workers are insufficient to affect the general pattern.

TABLE 4.1

VALUES OF k_p FOR MONOMERS AT 60°C

Monomer	k_p in mole^{-1} litre^{+1} sec^{-1}
styrene	176
methyl methacrylate	734
acrylonitrile	1960
methyl acrylate	2090
vinyl acetate	3700

Since the propagation reaction involves two reactants, it is not possible, from values of k_p alone, to decide whether the reactivity of the polymer radical or that of the monomer has the greater influence upon k_p. To distinguish between radical and monomer reactivities, it is necessary to consider the rates of transfer reactions and co-polymerizations. It is then possible to arrange monomers in order of reactivity towards particular reference radicals, and polymer radicals in order of reactivity towards reference transfer agents or reference monomers; the reference radicals and molecules must be picked carefully so that special polar effects on reactivity are absent (see Section B.*4*). Polymer radicals can be arranged thus in order of reactivity: polystyrene < polymethyl methacrylate < polyacrylonitrile < polymethyl acrylate < polyvinyl acetate. The order of reactivity of monomers is the reverse of this. From Table 4.1, it is evident that the reactivity of the polymer radical has a greater effect than that of the monomer upon the value of k_p.

The propagation reaction in the polymerization of an olefin corresponds to the conversion of a double bond into two single bonds and is strongly exothermic. The overall heat of polymerization is made up of contributions from all the elementary reactions but the largest by far is due to the propagation reaction if the conditions are such that high polymer is produced. In sensitized polymerizations, the heat of polymerization may increase slightly with rising [sensitizer] because of reactions involving the sensitizer and the primary radicals and also the termination processes. The additional contributions may represent about 5% of the total observed heat of polymerization; they can be eliminated by extrapolating to find the heat of polymerization corresponding to zero concentration of sensitizer (Dainton, Ivin and Walms ley, 1960). Heats of polymerization shown in Table 4.2 are taken from a collection by Flory (1953); some of the determinations were by direct calorimetry and others by comparison of the heats of combustion of monomer and polymer. Properly, the physical states of monomer and polymer should be defined but in most cases the differences between the

TABLE 4.2

HEATS OF POLYMERIZATION FOR OLEFINIC MONOMERS

Monomer	Heat of polymerization (kcal/mole)
vinyl acetate	21·3
methyl acrylate	18·7
acrylic acid	18·5
acrylonitrile	17·3
styrene	16·4
methacrylic acid	15·8
vinylidene chloride	14·4
methyl methacrylate	13·0
α-methyl styrene	9·0

various values are of minor importance. Certain heats of polymerization given by Dainton, Ivin and Walmsley (1960) differ slightly from those quoted here but not sufficiently to affect the general conclusions which can be drawn.

The observed heats of polymerization are almost all lower than the calculated values, the differences being particularly pronounced for the 1:1 disubstituted monomers. The calculated values are based on the assumption of no steric repulsion between groups within the polymer chains; it appears, therefore, that in certain polymers repulsions of this type must be significant.

Since polymerization is an exothermic process and its rate increases with rising temperature, there may be a thermal explosion if the heat is not dissipated sufficiently rapidly. Usually, however, the system reaches a steady state in which the temperature of the reaction mixture is higher than that of the medium surrounding the reaction vessel. If simplifying assumptions are made, the difference between the temperatures of the inner and outer walls of the reaction vessel can be calculated. Temperature effects may be quite important if polymerization is rapid, if large volumes of reactants are used so that the ratio of surface area to volume is small, and if the mixture becomes very viscous so that convection currents are slow.

Many of the methods developed for following the non-steady states in radical polymerizations, and used to evaluate kinetic constants, depend upon the small rise in temperature of the reaction system (Melville, 1956). If the reaction is photo-initiated, switching on the light causes the rate of initiation to change abruptly from zero to a finite value; the rate of polymerization builds up and the temperature rises. Reverse effects occur on switching off the light. If the system is almost adiabatic, changes in temperature can be used to follow the reaction during the build-up period; these changes can be followed directly by sensitive devices, such as thermocouples, or indirectly by measurements of dielectric constant or refractive index.

2. Effects of Temperature

Usually the growth or propagation reaction requires an activation energy of only about 5 kcal/mole, so that its rate does not vary very rapidly with temperature. Provided that radicals are generated at a suitable rate, radical polymerizations can be performed at quite low temperatures. Such reactions may be of considerable interest because:

(*a*) it may be possible to prepare truly linear polymers free from branches (see Section D of Chapter 5);

(*b*) stereo-regulated polymers may be produced (see Section A.*4*). Bovey (1960) has polymerized methyl methacrylate at $-78°C$ by a radical process using a source of γ-rays to generate radicals; the polymers were used in connection with a study of stereo-regulation in the growth process.

The kinetic chain length (ν) in a polymerization is given by

$$\nu = \frac{\text{rate of consumption of monomer}}{\text{rate of initiation}} = \frac{k_p[\text{P}\cdot][\text{M}]}{R_i} \tag{2}$$

If radicals are removed in pairs, in the stationary state

$$R_i = k_t[\text{P}\cdot]^2$$

and so

$$\nu = \frac{A_p}{A_t^{1/2}R_i^{1/2}}[\text{M}]\exp(-E_p+\tfrac{1}{2}E_t)/RT \tag{3}$$

Consider a series of polymerizations over a range of temperatures, [M] being fixed and R_i being kept constant by using photo-initiation. E_p is greater than E_t, so that, as the temperature is reduced, the kinetic chain becomes shorter. If the kinetic chain lengths are measured, it should be possible to determine $(E_p - \frac{1}{2}E_t)$ from a plot of $\ln \nu$ against (absolute temp.)$^{-1}$.

Consider now the relationship between the molecular chain length $(\bar{P})$ of the polymer, and the reaction temperature for the set of reactions specified above. If transfer is neglected, the limiting cases are

$$\bar{P} = 2\nu \quad \text{and} \quad \bar{P} = \nu$$

corresponding to exclusive combination and exclusive disproportionation respectively. If the mechanism of termination is unaffected by changes in temperature, the value of $\bar{P}$ should decrease as the temperature is reduced. Any shift of the balance between combination and disproportionation with temperature will almost certainly favour combination at low temperatures; the decrease of $\bar{P}$ with falling reaction temperature may therefore be a little less than expected from (3), but any such effect will certainly be small compared with one due to transfer reactions.

Transfer generally requires an activation energy larger than that for the growth process; its effect in reducing the average molecular weight of the polymer will therefore become more pronounced as the temperature is raised. If allowance is made for transfer, the molecular weight passes through a maximum as the temperature is raised. Below the temperature corresponding to this maximum, the molecular weight falls because the kinetic chain length decreases as the temperature is reduced; above this temperature, the molecular weight falls with increasing temperature because the increasing frequency of transfer reactions outweighs the effect due to the increase in the kinetic chain length. Burnett, George and Melville (1955) demonstrated this effect in the case of vinyl acetate, the maximum molecular weight being found for a polymer prepared at about 10°C.

The arguments developed above have been based on the assumption that the rate of initiation is independent of temperature. If radicals are generated by a thermal process requiring an activation energy E_i, (3) must be modified by replacing R_i by a term of the form $R_i^* \exp(-E_i/RT)$, where R_i^* is independent of temperature. The dependence of kinetic chain length upon temperature is then given by

$$\frac{d \ln \nu}{dT} = \frac{E_p - \frac{1}{2}E_t - \frac{1}{2}E_i}{RT^2} \tag{4}$$

If E_i is large, the kinetic chain length must decrease as the temperature of polymerization is raised; the effect will be accentuated by the greater frequency of transfer at higher temperatures, and also by the possibility of disproportionation becoming more significant at higher temperatures.

The treatment just given ignores the fact that at high temperatures the depropagation process

$$P_n\cdot \longrightarrow P_{n-1}\cdot + M \tag{5}$$

becomes significant. E_5 is likely to be about 20 kcal/mole; the rate of depropagation, therefore, increases very rapidly as the temperature becomes high. There is a temperature (T_c) at which the rate of depropagation becomes equal to that of propagation, so that

$$k_p[P\cdot][M] = k_5[P\cdot]$$

Expanding the velocity constants gives

$$A_p e^{-E_p/RT_c}[M] = A_5 e^{-E_5/RT_c}$$

whence

$$T_c = \frac{E_p - E_5}{R\ln(A_p[M]/A_5)} = \frac{\Delta H_p}{R\ln(A_p[M]/A_5)} \tag{6}$$

The temperature T_c is known as the *ceiling temperature* for the system (Dainton and Ivin, 1958).

The ceiling temperature can also be considered from a thermodynamic point of view. Although the difference between the heat contents of the initial and final states favours propagation, the reverse reaction is favoured by the entropy change. Polymerization is a process of association so that there is a decrease in entropy, but depropagation is accompanied by an increase in entropy since it is a dissociative process. At low temperatures, the free energy decreases during propagation because the heat content term outweighs the entropy term; above the ceiling temperature, however, $T\Delta S_p$ is numerically greater than ΔH_p, so that depropagation is favoured thermodynamically. At the ceiling temperature, ΔG is zero, so that

$$\Delta H_p = T_c \Delta S_p$$

and

$$T_c = \Delta H_p / \Delta S_p \tag{7}$$

Combining equations (6) and (7) gives

$$\Delta S_p = R\ln(A_p/A_5) + R\ln[M] = \Delta S_p^0 + R\ln[M]$$

where ΔS_p^0 is the entropy change accompanying polymerization at the standard state when the concentration (or properly, the activity) of monomer is unity; thus

$$T_c = \frac{\Delta H_p}{\Delta S_p^0 + R\ln[M]} \tag{8}$$

Since ΔH_p and ΔS_p^0 are negative, increasing [M] causes the ceiling temperature to rise. An alternative expression of this relationship is that, at any temperature, a high polymer must be in equilibrium with monomer at a concentration defined by equation (8). It must be emphasized that this is a thermodynamic approach; in the absence of active centres, a polymer appears stable even above the ceiling temperature, being in a state of meta-stable equilibrium. Another point of importance is that the expression for ceiling temperature is independent of the number or nature of the active centres in the system; for a given value of [M], the ceiling temperature should therefore be the same whether the active centres are radicals or ions.

For polymers derived from vinyl monomers, the concentrations of monomer in equilibrium with polymer are very low at ordinary temperatures; for example, at 25°C the calculated values for polystyrene and polymethyl methacrylate are respectively 10^{-6} mole/l. and 10^{-3} mole/l. For many systems, the equilibria cannot be established at higher temperatures because of side-reactions. The methacrylates, however, have comparatively low ceiling temperatures, chiefly because the heats of polymerization are rather low; at temperatures at which side-reactions are unimportant, there may be sufficient monomer in equilibrium with polymer to permit direct measurement. Such measurements with polymethyl methacrylate (Small, 1953) agreed with those made by a kinetic method (Ivin, 1955); polyethyl methacrylate also has been examined (Cook and Ivin, 1957).

Bywater (1955) studied the photo-sensitized polymerization of methyl methacrylate in solution over the range 100–150°C. The reactions did not proceed to completion; at each temperature, the final value of [monomer] was independent of the initial value and corresponded to the concentration of monomer in equilibrium with polymer. At 132·2°C, for example, the limiting value of [monomer] was close to 0·30 mole/l. (see Fig. 4.1).

If equation (8) is expressed in the form

$$\ln [\mathrm{M}] = \frac{\Delta H_p}{RT} - \frac{\Delta S_p^0}{R} \qquad (9)$$

it becomes apparent that measurements of the equilibrium concentration of monomer over a range of temperatures could lead to values for the changes in heat content, entropy and free energy accompanying polymerization. The various sets of data for methyl methacrylate give results in reasonable agreement with those found by other methods.

Schulz and Blaschke (1942) reported deviations from the expected kinetics for the polymerization of methyl methacrylate at temperatures above 125°C; they can be attributed to the depropagation process being significant under these conditions. For this monomer, Bywater (1955) showed that normal kinetic equations were satisfied if, instead of the actual concentration of monomer, the effective concentration, viz.

(actual concentration of monomer) – (final steady concentration),

was used. It was then possible to determine values for k_p at temperatures as high as 155°C, and to show that these values fitted with those determined at lower temperatures where depropagation is unimportant.

Effects of depropagation upon the kinetics of the overall polymerization are shown most clearly in the co-polymerization of alkenes with sulphur dioxide to give polysulphones (Dainton and Ivin, 1958). For

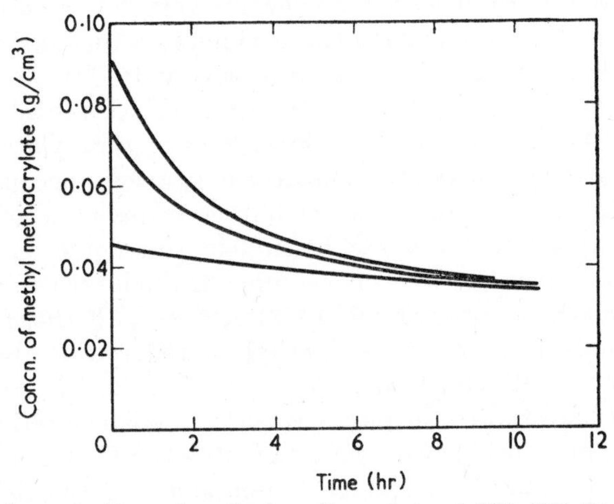

FIG. 4.1. Polymerization of methyl methacrylate at 132·2°C (Bywater, 1955).

some such systems the ceiling temperatures are below room temperature even for fairly high concentrations of monomers; experimental study of these systems is comparatively easy and there is little interference from side-reactions. For thermal initiation, the rate of co-polymerization rises with temperature, but passes through a maximum and falls to zero at the ceiling temperature; for photo-initiation, the maximum is not well developed, but as the ceiling temperature is approached the rate falls to zero.

In addition to decreasing the overall rate of polymerization, occurrence of depropagation upsets the normal variations of kinetic chain length and degree of polymerization with temperature. The kinetic chain length is the average *net* number of growth reactions in which an active centre participates during its life, and is given by

(No. of propagation steps) – (No. of depropagation steps).

Since the energy of activation for depropagation is greater than that for propagation, the second term becomes more important as the temperature rises; as the ceiling temperature is approached, the kinetic chain length, and also, therefore, the degree of polymerization of the resulting polymer, fall towards zero. This effect has been recorded by Dainton and Ivin (1952) for but-1-ene with sulphur dioxide, and provides another method for finding ceiling temperatures. In practice, values of the ceiling temperature from both the rate and molecular-weight methods may be slightly dependent upon the method of initiation, since the reactivity of a very small radical can be modified by changing its end-group.

Structural effects upon the polymerizability of compounds can be examined from the thermodynamic point of view. It is necessary to specify a standard state for all compounds to be considered, and then to compare either the changes in free energy accompanying polymerization at a given temperature, or to compare the ceiling temperatures. For derivatives of ethylene, the effect of structure upon ΔG_p is mainly due to differences between the various values of ΔH_p. The values of ΔS_p for most monomers are quite similar; thus the values, for specified states of monomer and polymer at $-20°C$, for styrene and α-methyl styrene are respectively -22 and -26 cal deg^{-1} mole^{-1} (Dainton and Ivin, 1958), although the monomers have very different ceiling temperatures.

Dainton and Ivin (1958) pointed out a fallacy in the common test for deciding whether a particular compound can be polymerized. Demonstration that a compound does not polymerize after prolonged heating in the presence of a source of radicals, is not proof that the compound cannot be polymerized by a radical mechanism since the test may have been conducted near, or above, the ceiling temperature. A much more reliable test is to examine the substance in a similar way at comparatively low temperatures. In this connection, Lowry (1958) has shown, contrary to earlier beliefs, that α-methyl styrene can be polymerized by a radical mechanism; the reaction is slow and it can be observed only at fairly low temperatures, since the ceiling temperature for the bulk monomer is in the region of 60°C. At 0°C the concentration of monomer in equilibrium with polymer is $0 \cdot 76$ mole/l. (Worsfold and Bywater, 1957), and even at $-40°C$ there are measurable amounts of monomer in equilibrium with polymer. The low ceiling temperature for this monomer is due to the fact that ΔH_p is only about -9 kcal/mole.

3. *Effects of Viscosity and Pressure*

The effect of the viscosity of the medium upon the diffusion of large radicals, and so upon the normal bimolecular termination process in radical polymerizations, is fully recognized (see Chapter 6, C). It is usually assumed that the rate of the growth reaction is hardly affected by changes in viscosity, since it involves the diffusion of the small monomer molecule; at very high viscosity, however, diffusion even of small molecules may be impeded, with a consequent reduction in the rate of propagation.

Bengough and Melville (1955) made a complete study of these effects for the polymerization of bulk vinyl acetate for conversions up to about 75%, at which point there was quite a rigid gel. The rate of polymerization increased until about 50% conversion, and then decreased quite rapidly with further reaction being very close to zero at 75% conversion. The polymerization was photo-initiated so that the velocity constants, k_p and k_t, could be determined at various conversions from examination of the non-stationary states; the activation energies also were evaluated.

TABLE 4.3

POLYMERIZATION OF VINYL ACETATE AT 25°C

% polymerization	k_p (mole^{-1} l.$^{+1}$ sec^{-1})	E_p (kcal/mole)
4	895	4·2
23	1290	4·2
46	1980	3·6
57	555	7·4
65	87	11·4

Too much reliance must not be placed on the precise numerical values for these quantities, if only because of doubt concerning the effective concentrations of monomer in extremely viscous mixtures, but the results in Table 4.3 show that in the later stages E_p increases quite considerably and k_p decreases.

A change in E_p from 4 to 7 kcal/mole would of itself reduce k_p by a factor of about 150; the observed effect is much less than this, and it appears that A_p also increases, partially compensating for the increase in E_p. The increasing viscosity of the medium could be responsible for the changes in both A and E; a high viscosity reduces the rate at which the two reactants come together, but it also reduces the rate at which they can separate. The time for which they may remain in close proximity is therefore increased, and there is a larger chance of them interacting during an encounter; a type of cage-effect can be considered. For vinyl acetate, the growth reaction becomes diffusion-controlled at about 50% conversion; the polymerization stops well before complete conversion simply because monomer cannot reach the trapped polymer radicals. For other monomers, diffusion-control of the growth reaction sets in at different stages in the polymerization. The presence of immobile radicals in gels has been confirmed by electron-spin resonance spectroscopy (Ingram, 1958).

It may not be valid to apply ordinary kinetic methods to evaluate the apparent velocity constants for the elementary reactions at all stages in polymerizations. At high conversions, there are significant deviations from the usual dependence of rate of polymerization upon rate of initiation. When termination is controlled by diffusion (and this may be at quite an early stage in the polymerization), the reaction rate increases with time, and the system is not in a stationary state because the rate of initiation exceeds the rate of termination.

Attempts to predict the effects of viscosity upon the elementary processes involved in polymerization (Vaughan, 1952; Robertson, 1956) are based upon an equation, due to Rabinowitch, relating the velocity constant for a bimolecular reaction to the diffusion coefficients of the reactants; the equation was derived for systems in which the reactants are identical with the solvent so that extension without modification can be regarded as approximate only. It is possible to estimate critical viscosities at which various reactions should become diffusion-controlled. The critical viscosity may be comparatively low when the activation energy for the actual reaction is low, and particles having low diffusion coefficients are involved; an example of such a reaction would be the bimolecular termination process in a radical polymerization. Propagation is a reaction of higher activation energy, and one of the reactants, being a small molecule, diffuses fairly readily; this reaction, therefore, can become diffusion-controlled only at higher viscosities. A still higher value for the critical viscosity would be expected for transfer to monomer, since the activation energy is higher than for propagation. The initiation process, in which a small radical reacts with monomer, involves particles which can diffuse quite readily, so that diffusion-control can set in only if the viscosity of the medium becomes very high. Vaughan's results for the elementary reactions in the polymerization of styrene at 125°C are summarized in Table 4.4.

The bulk viscosity for pure polystyrene, corresponding to 100% conversion, is about $1 \cdot 4 \times 10^9$ poises at 125°C for polymer of average molecular weight 360,000. From Table 4.4 it appears that termination, propagation and possibly also transfer to monomer might become diffusion-controlled during the bulk polymerization of this monomer at 125°C. The calculated critical viscosities are in general agreement with the viscosities of the reaction mixture when, according to kinetic evidence, termination and propagation are affected by viscosity.

It must be emphasized that the effects of the viscosity of the medium upon the velocity constants for the growth and other elementary reactions are not due to changes in the reactivities of the radicals and molecules. They arise from a purely physical effect, viz. that at high viscosities the controlling factor is the coming together of the reactants and not their chemical interaction.

TABLE 4.4

DIFFUSION-CONTROL OF ELEMENTARY REACTIONS FOR STYRENE

Reaction	Critical viscosity (poises)
Termination	2
Propagation	7×10^5
Transfer to monomer	8×10^8
Initiation	8×10^{16}

The effect of pressure upon the rate of a chemical reaction can be expressed by the equation

$$\frac{\mathrm{d} \ln k}{\mathrm{d}p} = -\frac{\Delta V^{\ddagger}}{RT} \tag{10}$$

where k = the velocity constant for the reaction,
$\Delta V^{\ddagger}$ = (partial molar volume of transition state) − (partial molar volume of reactants).

The growth reaction in polymerization is a process of association accompanied by a contraction; the transition state is intermediate between the reactants and the products, so that $\Delta V^{\ddagger}$ also is negative. Equation (10) predicts, therefore, that at constant temperature the velocity constant, k_p, would rise exponentially with pressure; inspection of Fig. 4.2 (Nicholson and Norrish, 1956) shows that this is approximately so for styrene at 30°C.

To consider the effects of pressure upon the overall rate of polymerization and the molecular weight of the resulting polymer, account must be taken of effects upon all the component reactions. For a polymerization in which radicals are both generated and destroyed in pairs, the effect of pressure upon the overall rate can be expressed as

$$\frac{\mathrm{d} \ln k}{\mathrm{d}p} = \frac{-(\Delta V_p^{\ddagger} + \frac{1}{2}\Delta V_i^{\ddagger} - \frac{1}{2}\Delta V_t^{\ddagger})}{RT} \tag{11}$$

where $k =$ the composite velocity constant for the overall reaction, and $\Delta V_p^{\ddagger}$, $\Delta V_i^{\ddagger}$ and $\Delta V_t^{\ddagger}$ refer to the differences between the partial molar volumes of transition state and reactants for propagation, dissociation of initiator and termination respectively. The overall rate of polymerization of styrene at 60°C increases continuously with pressure, at least up to 5000 kg/cm^2, but the relationship is a little more complicated than indicated by (11) (Merrett and Norrish, 1951).

Since the rate of polymerization goes up as the pressure increases, and at the same time the rate of production of radicals from the initiator falls slightly, the kinetic chain length must rise with increasing pressure; accordingly the average molecular weight of the polymer should rise with the pressure. Merrett and Norrish, however, showed that the molecular

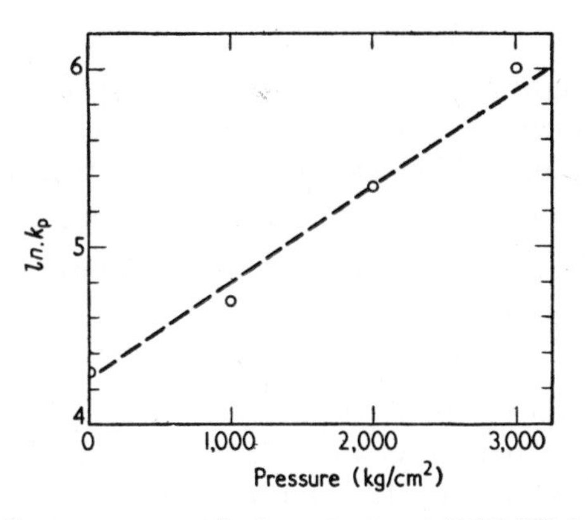

FIG. 4.2. Effect of pressure upon k_p for sytrene at 30°C (Nicholson and Norrish, 1956).

weight tended to a limiting value; this was attributed to increased frequency of transfer reactions at very high pressures, and also to the occurrence under these conditions of a termination process involving a polymer radical and a molecule of the initiator.

4. *Mechanism*

Four possible growth reactions must be considered, viz.

$$P.CH_2.CXY\cdot + CH_2{:}CXY \longrightarrow P.CH_2.CXY.CH_2.CXY\cdot \quad (12)$$
$$P.CH_2.CXY\cdot + CXY{:}CH_2 \longrightarrow P.CH_2.CXY.CXY.CH_2\cdot \quad (13)$$
$$P.CXY.CH_2\cdot + CXY{:}CH_2 \longrightarrow P.CXY.CH_2.CXY.CH_2\cdot \quad (14)$$
$$P.CXY.CH_2\cdot + CH_2{:}CXY \longrightarrow P.CXY.CH_2.CH_2.CXY\cdot \quad (15)$$

Repetition of (12) would lead to polymer molecules of regular structure having substituents on alternate carbon atoms of the main chain; this is known as the head-to-tail or 1,3-structure. The other simple regular structure for the polymer chain is

$$\cdot CH_2.CXY.CXY.CH_2.CH_2.CXY\cdot$$

resulting from alternate additions according to (13) and (15). There is considerable chemical and physical evidence that the 1,3-structure predominates in vinyl polymers (Flory, 1953), and that it results from repetition of (12) rather than of (14); in the growing polymer radical, therefore, the unpaired electron is associated with the carbon atom carrying the substituents.

Although 1,3-addition predominates, occasional head-to-head and tail-to-tail additions may occur. Further, if polymer radicals finally terminate by combination, head-to-head linkages are produced by the reaction

$$2P.CH_2.CXY\cdot \longrightarrow P.CH_2.CXY.CXY.CH_2.P \tag{16}$$

The molecular weight of polyvinyl alcohol is reduced by treatment with periodic acid (Flory and Leutner, 1948), suggesting the presence of 1,2-diglycol groupings which are cleaved by the reagent

$$P.CH_2.CH(OH).CH(OH).CH_2.P \longrightarrow 2P.CH_2.CHO \tag{17}$$

From the fall in average molecular weight, the fraction (δ) of 1,2-linkages in the chain is given by

$$\delta = 0 \cdot 10 \exp(-1 \cdot 3\,\text{kcal}/RT)$$

where T is the temperature of preparation of the polyvinyl acetate used as the source of the polyvinyl alcohol. The fraction is governed by the ratio of the rates of (13) and (12), and $(E_{13}-E_{12})$ is 1·3 kcal/mole.

Comparison of the energetics of the various alternative reactions supports the view that (12) is the main propagation step. The radical produced in (12) is more stable than that formed in (14); this is due to the fact that for the former it is usually possible to write alternative structures in which the unpaired electron is sited on a substituent. This effect is particularly pronounced for monomers containing the phenyl group; for the polystyrene radical, for example, the following structures may be considered:

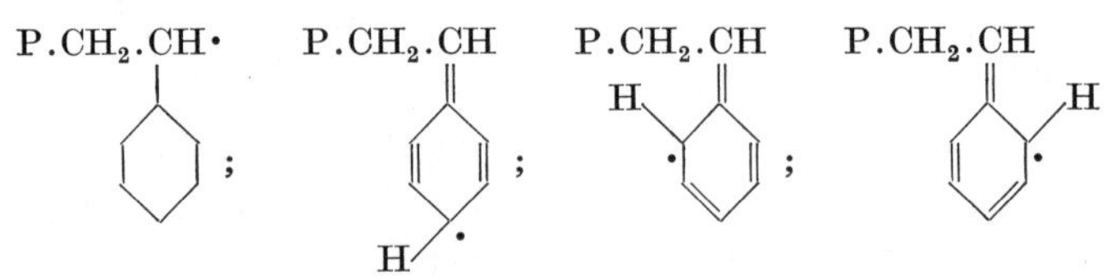

For reasons discussed in Section B.*3*, this stabilization of the product is likely to make E_{12} less than E_{13} and E_{14}; E_{15} also would be comparatively small, but (15) cannot be the sole growth reaction since it does not regenerate the original radical, and it would have to alternate with (13). For styrene, $(E_{13}-E_{12})$ is probably about 9 kcal/mole; as shown already, the difference in the case of vinyl acetate is only about 1·3 kcal/mole, because resonance stabilization due to the acetate group is only slight. The high reactivity of the polyvinyl acetate radical and the large value of k_p for this monomer are due to the same cause; it can be predicted, therefore, that head-to-head additions are rather more likely for monomers having high values for k_p.

Electron spin resonance studies can provide information about the structures of radicals, and the technique has been applied to growing polymer radicals (Ingram, 1958; Whiffen, 1959). In order to raise the concentrations of radicals to suitable levels, it is necessary to examine systems in which radicals become trapped in precipitates, glasses or gels. It has been concluded that the polymethyl methacrylate radicals have the structure

$$P.CH_2.C(CH_3)(COOCH_3)\cdot$$

in accord with head-to-tail addition by (12).

Solid polymethyl methacrylate after exposure to high-energy radiations exhibits an electron spin resonance spectrum identical with that of the radicals trapped in gelled co-polymers of methyl methacrylate and glycol dimethacrylate, and corresponding to a single species of radical. Main-chain scission of the polymer is believed to occur during the irradiation, and it must give rise to two distinct types of radical, viz.

$$P.CH_2.C(CH_3)(COOCH_3).CH_2.C(CH_3)(COOCH_3)\cdot$$
$$\text{and}\quad P.C(CH_3)(COOCH_3).CH_2.C(CH_3)(COOCH_3).CH_2\cdot$$

Failure to detect the second type of radical has been accounted for by the greater stability of radicals of the first type (Ingram, Symons and Townsend, 1958). The radicals may shed monomer molecules as a result of the excess energy in the system but subsequently re-polymerization occurs. The structure of the first type of radical is unaffected by this process but that of the second type is changed, thus:

$$P.C(CH_3)(COOCH_3).CH_2.C(CH_3)(COOCH_3).CH_2\cdot$$
$$\longrightarrow P.C(CH_3)(COOCH_3).CH_2\cdot + C(CH_3)(COOCH_3){:}CH_2$$
$$\longrightarrow P.C(CH_3)(COOCH_3).CH_2.CH_2.C(CH_3)(COOCH_3)\cdot \quad (18)$$

As a result, only a single species of radical is preserved in the polymer.

In polymers having the repeating unit $\cdot CH_2.CXY\cdot$, considerable interference between the pendant groups is revealed during attempts to make molecular models. In these polymers tail-to-tail arrangement of monomer units is the most favoured sterically since the substituents are kept well apart, but, as explained already, (15) cannot be the sole growth reaction. The heats of polymerization for disubstituted monomers such as methyl methacrylate are appreciably lower than those for monosubstituted monomers as a result of this steric effect (see Table 4.2).

Quite apart from head-to-head addition, other types of abnormal growth reactions are possible in certain cases. Although they occur infrequently, they cannot be neglected because they give rise to unusual groupings in the polymer chains, and these may act as centres for degradation, branching, and other types of reaction. Some examples of systems in which alternative propagation reactions may occur are discussed below.

Talât-Erben and Bywater (1955*a*) showed, from spectroscopic evidence, that polymethacrylonitrile prepared by radical reactions contains a few ketene-imine groups which can arise only if the polymer radical occasionally reacts in the form

$$P.CH_2.C(CH_3):C:N\cdot \quad \text{instead of} \quad P.CH_2.C(CH_3)(CN)\cdot$$

A similar effect has been discussed (see Chapter 2, C) for the 2-cyano-2-propyl radical. The ketene-imine group is not stable, and its characteristic absorption peak disappears on heating the polymer. There is no evidence for ketene-imine groups in polyacrylonitrile as ordinarily prepared but they are present in polymers prepared at −78°C using X-ray initiation (Chen, Colthup, Deichert and Webb, 1960). Other polymer radicals also may react in abnormal forms, for example, polymethyl methacrylate as $P.CH_2.C(CH_3):C(OCH_3).O\cdot$. There is very little direct evidence for the presence of abnormal units in the polymer; it is significant, however, that the radical $(CH_3)_2C(COOCH_3)\cdot$, which can be regarded as a model for the polymethyl methacrylate radical (see Chapter 2, C), may react in an alternative form.

In the cases just discussed, the polymer radical is considered as reacting in an abnormal form, but it is conceivable that the monomer might do so also. For monomers containing the benzene ring, it is necessary to consider the possibility of radical attack upon the ring. In the case of styrene, this is of very little importance; for vinyl benzoate and related monomers, however, it is thought (Ham and Ringwald, 1952) that the normal growth reaction may be accompanied by

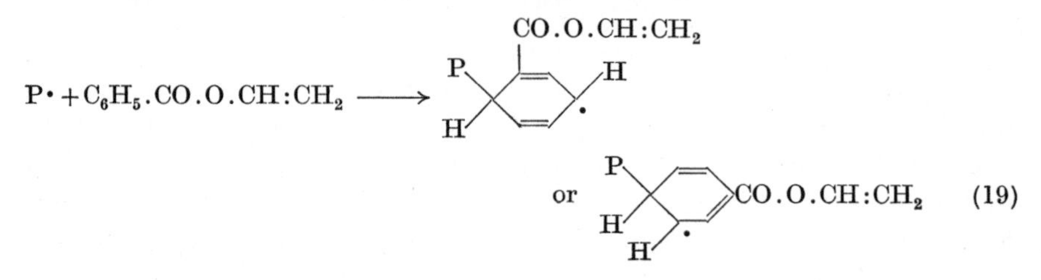

The unreacted $\cdot CH:CH_2$ groups are subsequently attacked by growing polymer radicals eventually giving network structures. The stabilized radical formed in (19) is relatively unreactive and is responsible for "retardation by monomer" in the polymerization (Litt and Stannett, (1960).

Another distinct case of alternative propagation reactions is found in polymerizations of ionizable monomers in ionizing solvents such as water; four growth reactions must be considered, viz.

$$P\cdot + M \longrightarrow P\cdot \qquad (20)$$

$$P\cdot^* + M \longrightarrow P\cdot \qquad (21)$$

$$P\cdot + M^* \longrightarrow P\cdot^* \qquad (22)$$

$$P\cdot^* + M^* \longrightarrow P\cdot^* \qquad (23)$$

where M* represents an ionized monomer molecule and P·* a polymer radical in which the terminal monomer unit is ionized. Reaction (23) would be expected to be the slowest of the four because of electrostatic repulsion between the like charges of M* and P·*. In a complete treatment, account would be taken of effects upon reactivity brought about by ionization of non-terminal monomer units.

The polymerization of methacrylic acid in water has been studied (Katchalsky and Blauer, 1951; Pinner, 1952) at various pH's, so that it was possible to vary the proportions of monomer molecules and polymer radicals in the ionized forms. Variation of rate of polymerization with pH could be attributed in part to differences between the rates of the various growth reactions, and in part to differences between the rates of the three types of interaction between P· and P·* radicals. Polymerizations in aqueous solutions are difficult experimentally, the rate of polymerization being very sensitive to trace impurities, particularly oxygen and heavy metal ions. The results displayed in Fig. 4.3 show clearly, in spite of scatter, that the rate of polymerization decreases as the pH rises, i.e. as the degrees of ionization of monomer and radical increase. Between pH 6 and 11, the rate rises slowly (Blauer, 1953, 1960).

The degree of ionization of the monomer at any pH can be calculated using the value for the dissociation constant, but similar calculations for the polymer radical cannot be made with certainty because of doubt concerning the dissociation constant; Pinner considered that the dissociation constants for monomer and polymer radical can be taken as equal, while Katchalsky and Blauer took the constant for the radical as equal to that for the dead polymer which is less than that for the monomer. If the monomer is a stronger acid than the polymer, two points must be considered:

(*a*) in an unbuffered medium, the pH must rise during polymerization with a consequent fall in the rate of polymerization;

(*b*) an ionized polymer radical produced in (22) is not in electrochemical equilibrium with its environment.

The average interval between successive additions to a polymer radical is such that there is ample time for the ionized radical P·* to be converted to the unionized form P·; the ratios [M]/[M*] and [P·]/[P·*], therefore, remain fixed at values governed by the dissociation constants of the monomer and the polymer radical and by the pH of the medium. The polymerizations of ionizable monomers can therefore be regarded as co-polymerizations of a special type.

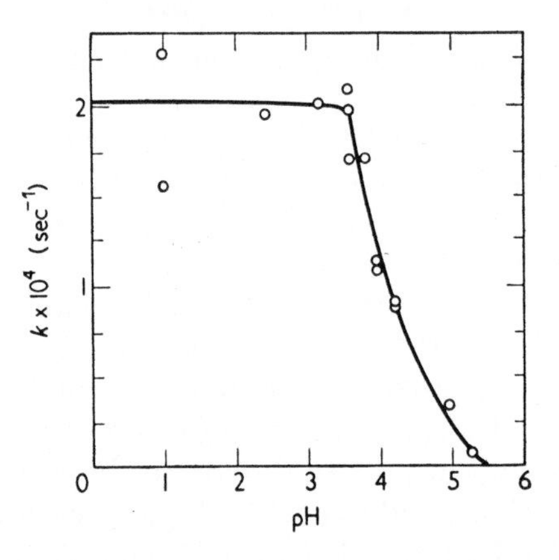

FIG. 4.3. Effect of pH upon rate of polymerization of methacrylic acid (0·25 M in water) at 75°C (Katchalsky and Blauer, 1951).

In the growth of a polymer chain, the head-to-tail addition of monomer to a reactive centre may occur in two ways, giving products which are distinct stereochemically. In a polymer having the repeating unit $\cdot CH_2 . CXY \cdot$, in which either X or Y may be hydrogen, the substituted carbon atoms are asymmetric; the attachment of another monomer unit to a centre causes the additional asymmetric carbon atom to acquire either a *d* or an *l* configuration. Ordinarily in radical polymerizations, the two types of growth reaction occur at random, so that the resulting polymers, referred to as "atactic", can be considered as special types of random co-polymers. A succession of placements of the same configuration gives an "isotactic" polymer molecule represented thus:

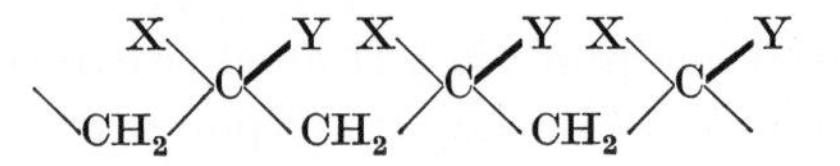

Alternate *d* and *l* placements give a "syndiotactic" polymer represented as

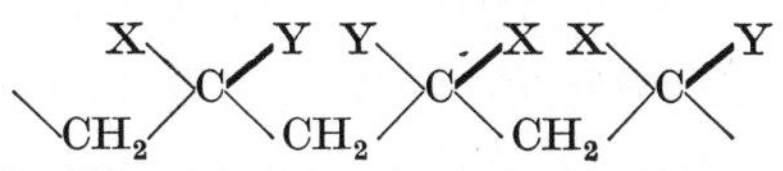

Isotactic and syndiotactic polymers are produced in "stereo-regulated" polymerizations; for the most part, these reactions occur by ionic mechanisms, depending upon the use of catalysts of special types, but there are reports that stereo-regulation can be present in certain radical polymerizations also. It may become necessary, therefore, in radical polymerizations, to consider two distinct values for k_p corresponding to the two types of placement.

The various types of polymer may differ in chemical reactivity; for example, isotactic polymethyl methacrylate is hydrolysed much more readily than the syndiotactic or atactic material (Glavis, 1959). The main difference between the various types, however, lies in the greater crystallizability of the isotactic and syndiotactic materials. The crystallizability of polymethyl methacrylate, prepared homogeneously with a photo-sensitizer, increases as the temperature of preparation is reduced; this suggests that at low temperatures there is some regulation of the growth reaction, and that the isotactic and syndiotactic placements have slightly different characteristics. It has been concluded that for methyl methacrylate, one type of placement may be favoured by 2 kcal/mole in the energy term, and opposed by a difference of 5 entropy units in the entropy of activation (Coleman, 1958; Fox, Goode, Gratch, Huggett, Kincaid, Spell and Stroupe, 1958). Bovey (1960) has examined samples of polymethyl methacrylate prepared by radical reactions at temperatures between $-78°C$ and 100°C and has deduced that the activation energy for an isotactic placement is 775 cal/mole greater than that for a syndiotactic placement and that the entropies of activation for the two types of growth reaction are equal. Even a small difference such as this would cause about 75% of the growth reactions at 100°C to lead to syndiotactic placement.

Examination of the infra-red spectra of samples of polyvinyl chloride prepared at temperatures between $-80°C$ and $+45°C$ (Grisenthwaite and Hunter, 1958) suggests that as the preparation temperature is reduced, structural irregularities in the polymer become less abundant. The effect is too large to be associated with chain-branching or with head-to-head and tail-to-tail additions, and it may be due to the presence of stereo-regularity in segments of the polymer chain.

B. Co-polymerization

1. Compositions of Products and Rates of Reaction

Co-polymers are of great technical importance, and the reactions by which they are produced have been extensively studied academically. In particular, attempts to elucidate the factors which govern the rates of co-polymerizations and the compositions of the products have led to much valuable information on the effects of the structures of the reactants upon the rates of radical reactions.

The composition of the co-polymer formed from a particular mixture of monomers depends upon the mechanism of the polymerization, i.e. whether radical, anionic or cationic; the effect is pronounced for styrene with methyl methacrylate (Landler, 1952). In radical reactions, a 1:1 mixture of these monomers gives a product with approximately equal numbers of the two types of monomer unit; in cationic reactions, the polymer is almost pure polystyrene, while in anionic reactions hardly any styrene is incorporated in the product. Examination of the composition of the product has been used in identification of the type of polymerization promoted by new and unusual initiators (see Chapter 2,D).

There are many types of binary co-polymer differing in distribution of monomer units in the polymer molecule. At the one extreme, the two types of unit may be arranged purely at random, but frequently they tend to be arranged alternately along the polymer chain. In certain systems alternation is complete, so that the unit in the chain is effectively a group consisting of the two monomer units $\cdot M_a . M_b \cdot$ and in some cases the growth reaction has been thought to involve a 1:1 complex of the two monomers. Another type of co-polymer is one in which long sections of the polymer molecule consist of the same type of monomer unit; in this case the molecule might be represented as

$$\cdot(M_a)_p . (M_b)_q . (M_a)_r . (M_b)_s \cdot$$

where p, q, r and s are quite large. Co-polymers of this type are called *block co-polymers*; they are not produced in any ordinary co-polymerizations. Related to the block co-polymers are *graft co-polymers* in which polymer chains of one type are attached as long branches to the chain of another polymer. Block and graft co-polymers are of great interest technically. Ingenious methods have been devised to prepare such materials free from homo-polymers and other types of co-polymer, and to characterize them (Smets, 1957).

In the kinetic analyses of radical co-polymerizations, it is usual to make the simplifying assumption that the reactivity of a polymer radical is governed entirely by the monomer unit last added; thus the radical $P.CH_2.CH(C_6H_5)\cdot$ is referred to as a polystyrene radical no matter what the nature of P. In certain systems (see Section B.*6*) the assumption is inadequate, but usually it leads to satisfactory prediction of the behaviour of mixtures of monomer.

On this assumption, the co-polymerization of a pair of monomers involves two types of polymer radicals and four distinct growth reactions:

$$P.M_a\cdot + M_a \longrightarrow P.M_a.M_a\cdot \quad \text{velocity constant} = k_{aa} \tag{24}$$
$$P.M_a\cdot + M_b \longrightarrow P.M_a.M_b\cdot \quad k_{ab} \tag{25}$$
$$P.M_b\cdot + M_a \longrightarrow P.M_b.M_a\cdot \quad k_{ba} \tag{26}$$
$$P.M_b\cdot + M_b \longrightarrow P.M_b.M_b\cdot \quad k_{bb} \tag{27}$$

The ratios k_{aa}/k_{ab} and k_{bb}/k_{ba} are defined as the *monomer reactivity ratios*, r_a and r_b. In a binary co-polymerization, there are four transfer-to-monomer reactions; if solvent is present, there are two types of transfer-to-solvent. Mutual termination can result from three types of radical interaction.

When dealing with the composition of a co-polymer, it is necessary only to consider the growth reactions, unless the average chain is so short that end-groups make significant contributions to the composition. It can then be derived that

$$\left[\frac{M_a}{M_b}\right]_{\text{co-polymer}} = \frac{[M_a]}{[M_b]}\cdot\frac{r_a[M_a]+[M_b]}{r_b[M_b]+[M_a]} \tag{28}$$

where $[M_a]$ and $[M_b]$ are the concentrations of the two monomers in the reaction mixture, and $[M_a/M_b]_{\text{co-polymer}}$ is the ratio of the numbers of the two types of monomer unit in the product.

If there are more than two monomers involved in a co-polymerization, the number of elementary reactions becomes large; for a system containing n monomers, there are n^2 growth reactions. Kinetic analysis shows that the compositions of polycomponent co-polymers could be predicted from the monomer reactivity ratios applicable to the appropriate binary co-polymerizations; this was confirmed (Walling and Briggs, 1945) for various combinations of styrene, methyl methacrylate, acrylonitrile and vinylidene chloride.

There are many examples of co-polymerizations involving monomers which do not undergo homo-polymerization; for these reactions, either or both the velocity constants k_{aa} and k_{bb} are zero but the cross-propagations can occur readily. Monomers exhibiting this behaviour include some 1,2-disubstituted ethylenes such as maleic anhydride, esters of maleic and fumaric acids, stilbene and cinnamic acid. Examples of other compounds which enter co-polymerizations are given in Table 4.5. Some of the reactions are accompanied by pronounced retardation, and the substances are better considered as retarders than as co-monomers (see Section D of Chapter 7).

Certain photo-excited molecules, e.g. anthracene, can act as co-monomers. Ordinarily anthracene co-polymerizes only to a small extent

TABLE 4.5

EXAMPLES OF UNUSUAL MONOMERS WHICH ENGAGE IN CO-POLYMERIZATION

Monomer	Monomer unit in co-polymer
sulphur dioxide	$-\mathrm{S(=O)_2}-$
carbon monoxide	$-\mathrm{C(=O)}-$
oxygen	$-\mathrm{O}-\mathrm{O}-$
p.benzoquinone	$-\mathrm{O}-\mathrm{C_6H_4}-\mathrm{O}-$
benzene	$-\mathrm{CH}(\mathrm{C_4H_4})\mathrm{CH}-$
cyclic disulphides	$-\mathrm{S}-[\mathrm{CH_2}]_n-\mathrm{S}-$

with styrene (Marvel and Wilson, 1958), the reaction being rather similar to that involved in the co-polymerization of benzene with vinyl acetate.

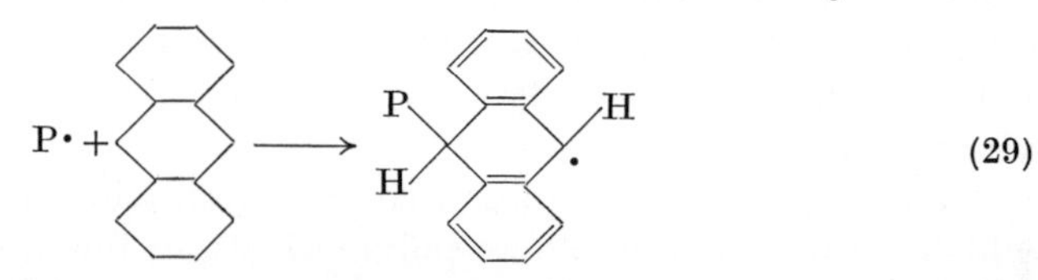

(29)

The co-polymerization is accompanied by retardation, and, for some monomers other than styrene, anthracene is an inhibitor. Evidently the product of (29) is rather unreactive because of resonance stabilization. Norrish and Simons (1959) showed that the polystyrene radical reacts extremely readily with triplet anthracene, the reaction requiring no activation energy. Relative values for the velocity constants at 50°C for the propagation reactions in the co-polymerization of styrene and anthracene are shown in Table 4.6.

Photo-excited states of benzoin and its methyl ether, but not the ground states, are probably capable of participating in co-polymerizations. When the methyl ether of benzoin is used as a photo-sensitizer for the polymerization of methyl methacrylate, about twelve molecules of the ether are incorporated in each polymer molecule (Mochel, Crandall and Peterson, 1955); a similar result is found when benzoin is used with styrene or methyl methacrylate (Bevington and Lewis, 1960*b*). The polymerizations have the characteristics of mono-radical reactions, and

TABLE 4.6

GROWTH REACTIONS IN THE CO-POLYMERIZATION OF STYRENE AND ANTHRACENE

Reaction	Relative velocity constant
$P\cdot + M \longrightarrow P\cdot$	1
$P\cdot + A \longrightarrow P.A\cdot$	3
$P\cdot + {}^3A \longrightarrow P.A\cdot$	$1\cdot1 \times 10^{9}$
$P.A\cdot + M \longrightarrow P\cdot$	$6\cdot7 \times 10^{-3}$

$P\cdot$ = polystyrene radical
M = monomeric styrene
A = anthracene
${}^3A$ = triplet anthracene

neither benzoin nor its ether enters polymers to any significant extent when the reactions are initiated with thermal sensitizers.

Generally the ratios of monomer concentrations in the feed and the product are not equal; as polymerization proceeds, the composition of the feed gradually changes and the average composition of the co-polymer drifts. Equation (28) refers to the co-polymer formed from a particular reaction mixture; if there is marked preferential consumption of one of the monomers, corrections are necessary. The changes in feed composition and therefore in co-polymer composition may be of great significance in commercial practice, since in high conversion co-polymers there may be considerable heterogeneity of composition. Integration of the instantaneous co-polymer composition equation (28) is possible, but for details, the relevant sections in the books by Bamford, Barb, Jenkins and Onyon (1958) and Alfrey, Bohrer and Mark (1952) should be consulted.

If two co-polymerizations are performed with different values of $[M_a]/[M_b]$ and the values of $[M_a/M_b]_{\text{co-polymer}}$ are determined, r_a and r_b can be calculated. The compositions of the co-polymers are usually determined by elementary analysis; in many cases this is unreliable or even impossible because of close similarity of the two monomer units. Even in the most favourable cases, difficulties may arise if in the co-polymer one type of unit is much less abundant than the other. Alternative methods, including analysis for functional groups and spectroscopic and tracer methods, have been applied to some systems. In practice, a large number of co-polymerizations must be performed in order to obtain reliable values for monomer reactivity ratios. Procedures for dealing with sets of experimental data have been fully described in the book by Alfrey *et al.* These authors have collected results for many monomer pairs, and more recent data have been tabulated by Walling (1957).

The rate of a co-polymerization, and also the average molecular weight of the product, can be expressed in terms of the kinetic constants of the elementary reactions, and the concentrations of the monomers. The treatment follows that for the polymerizations of single monomers in that it is normally assumed that

(*a*) radical reactivity is independent of radical size,
(*b*) a stationary state is set up, and
(*c*) the only reactions in which there is appreciable consumption of monomer are the growth reactions.

For binary co-polymerizations, the following additional assumptions are usually made:

(*d*) the rate of initiation is independent of the composition of the feed,
(*e*) any transfer reactions are not accompanied by retardation, and
(*f*) there are four growth and three bimolecular termination reactions.

It can then be shown that

rate of consumption of monomer =

$$\frac{R_i^{1/2}(r_a[M_a]^2+2[M_a][M_b]+r_b[M_b]^2)}{(r_a^2\delta_a^2[M_a]^2+2\phi r_a r_b\delta_a\delta_b[M_a][M_b]+r_b^2\delta_b^2[M_b]^2)^{1/2}} \tag{30}$$

where R_i = rate of initiation,

r_a and r_b = monomer reactivity ratios,

$[M_a]$ and $[M_b]$ = concentrations of monomers in the feed,

δ_a and δ_b = $k_{t_a}^{1/2}/k_{aa}$ and $k_{t_b}^{1/2}/k_{bb}$ respectively,

and $\phi = k_{t_{ab}}/(k_{t_a}.k_{t_b})^{1/2}$.

Equation (30) refers to an instantaneous rate; allowance can be made for the gradual change in rate resulting from a drift in composition of the feed, but the equation then becomes difficult. The chief interest in connection with the actual rates of co-polymerizations is in the term ϕ concerned with the termination reactions (see Section D of Chapter 6).

2. Distributions in Co-polymers

In a co-polymer, just as in a homo-polymer, there is a distribution of molecular sizes; the mathematical form of this distribution can be worked out if the detailed mechanism of the reaction is known. The value of the average chain length depends upon the composition of the feed, which may affect also the form of the distribution curve through its effect on the frequency of transfer reactions, and even perhaps on the mechanism of the termination process. The gradual change in composition of the feed during reaction will of itself, therefore, cause a shift in the distribution curve; consequently a co-polymerization allowed to proceed to an appreciable conversion gives a product having a distribution of sizes wider than that of a homo-polymer produced in a comparable reaction.

It is necessary also to consider variation of chemical composition between molecules in a sample of co-polymer. Stockmayer (1945) showed that for low-conversion co-polymers, the distribution of composition about the mean value can be represented by a Gaussian curve with a very sharp peak. In high-conversion materials, the distribution may be much wider.

In co-polymerizations for which both r_a and r_b are zero, the two types of monomer unit are arranged strictly alternately; the co-polymerization of maleic anhydride and stilbene approximates quite well to this, r_a and r_b both being $0{\cdot}03 \pm 0{\cdot}03$ (Lewis and Mayo, 1948). It is generally found that a tendency for alternation is superimposed on a random arrangement of monomer units. It is quite easy to calculate the frequencies at which particular numbers of units of the same type occur at adjacent sites in co-polymer chains (Alfrey, Bohrer and Mark, 1952); the calculation involves knowledge of the composition of the feed and of the monomer reactivity ratios. Unfortunately, there are no satisfactory methods for experimental tests.

3. Relative Reactivities of Monomers and of Radicals

The value of k_p for the polymerization of a monomer depends on the reactivities of both the monomer and the polymer radical derived from it. Even if absolutely reliable values for k_p for the various monomers were available, it would still not be possible to use them on their own for comparisons of the reactivities of the monomers, or of the corresponding radicals, since the effects of the monomer and the radical cannot be separated. At first sight, studies of co-polymerizations would only introduce further complications, but in fact much additional information can be obtained.

From the monomer reactivity ratios and the values of k_{aa} and k_{bb}, found from experiments with single monomers, the velocity constants for the cross-propagations can be evaluated. Suppose that several monomers are co-polymerized with a particular monomer, described as M_a, and that the various values of k_{ab} are determined. These values should be measures of the relative reactivities of the monomers towards a reference radical, viz. the polymer radical derived from M_a. Table 4.7 gives results for a few monomers with three reference radicals at 60°C;

TABLE 4.7

RELATIVE REACTIVITIES OF MONOMERS

	Relative velocity constant for reaction with		
Monomer	polystyrene radical	polymethyl methacrylate radical	polyvinyl acetate radical
styrene	1·00	2·18	100
methyl methacrylate	1·92	1·00	67
acrylonitrile	2·44	0·74	25
vinyl chloride	0·06	0·08	3
vinyl acetate	0·02	0·05	1

much more comprehensive tables have been given by Walling (1957) among others. Each column in Table 4.7 must be considered on its own since the velocity constants are expressed relative to that for the reaction of the radical with its own monomer.

Certain regularities can be recognized in these results; for example, the molecules of vinyl chloride and vinyl acetate are rather unreactive towards all three polymer radicals. Marked anomalies, such as the high reactivity of monomeric acrylonitrile towards the polystyrene radical, are associated with pairs of monomers having pronounced tendencies for alternation in co-polymerization.

Essentially, this method for studying the relative reactivities of monomers towards reference radicals is a competitive method of the type discussed in Chapter 3, D.*2*. There is competition between the two monomers for reacting with a polymer radical, and the rates of the alternative growth reactions are compared by analysing the product, in this case by determining the relative numbers of the two types of monomer unit in the co-polymer.

Examination of certain special ternary co-polymerizations can lead quite simply to comparisons of the reactivities of monomers towards reference polymer radicals (Walling, Seymour and Wolfstirn, 1948). In a system containing maleic anhydride (M_a) and two ring-substituted α-methylstyrenes (M_b and M_c), only four of the nine possible growth reactions occur to any appreciable extent at the working temperature; they are:

$$P.M_a\cdot + M_b \longrightarrow P.M_a.M_b\cdot \quad \text{velocity constant} = k_{ab} \qquad (31)$$

$$P.M_a\cdot + M_c \longrightarrow P.M_a.M_c\cdot \quad k_{ac} \qquad (32)$$

$$P.M_b\cdot + M_a \longrightarrow P.M_b.M_a\cdot \quad k_{ba} \qquad (33)$$

$$P.M_c\cdot + M_a \longrightarrow P.M_c.M_a\cdot \quad k_{ca} \qquad (34)$$

Appreciable quantities of the substituted styrenes are consumed only in the growth reactions (31) and (32), so that

$$-\frac{d[M_b]}{dt} = k_{ab}[P.M_a\cdot][M_b] \quad \text{and} \quad -\frac{d[M_c]}{dt} = k_{ac}[P.M_a\cdot][M_c]$$

and

$$\frac{d[M_b]}{d[M_c]} = \frac{k_{ab}[M_b]}{k_{ac}[M_c]}$$

By integration,

$$\log\frac{[M_b]_0}{[M_b]} = \frac{k_{ab}}{k_{ac}}\log\frac{[M_c]_0}{[M_c]} \qquad (35)$$

where $[M_b]_0$ and $[M_c]_0$ refer to initial concentrations. The ratio k_{ab}/k_{ac} can be found from a single experiment in which the ratio of the initial concentrations of the two styrenes is known and the ratio of the final concentrations of these monomers is determined. It is convenient to take unsubstituted α-methylstyrene as standard, and then the velocity constants for reactions of substituted α-methylstyrenes with the maleic anhydride radical can be expressed in terms of that for the standard. This is a competitive method for studying the reactivities of the monomers; essentially, product analysis is used for comparing the numbers of molecules of the two styrene derivatives captured by the maleic anhydride radical.

This method could be applied to other sets of monomers, chosen so that:

(*a*) none of them can undergo homo-polymerization,

(*b*) the monomer common to all experiments, and related to the reference radical, is able to enter binary co-polymers with all the other monomers,

(c) the other monomers cannot undergo binary co-polymerizations among themselves.

Monomer reactivity ratios can also be used to study the reactivities of polymer radicals towards reference monomers by comparing velocity constants k_{aa}, k_{ba}, k_{ca}, etc. For this purpose, it is essential to use absolute values of k_p for the homo-polymerizations of the various reference monomers. In constructing Table 4.8, use was made of values for k_p collected in Table 4.1, and monomer reactivity ratios listed in the book by Alfrey *et al.* (1952). The precise numerical values in this table may be suspect, but clearly the polyvinyl acetate radical is by far the most reactive, and the polystyrene radical the least reactive of the three. More extensive tables confirm that there is a general inverse relationship between the reactivity of a monomer and that of the corresponding polymer radical. It has been pointed out already that the magnitude of k_p for a monomer is influenced more by the reactivity of the polymer radical than by that of the monomer molecule.

TABLE 4.8

RELATIVE REACTIVITIES OF POLYMER RADICALS AT 60°C

Polymer radical	Velocity constant, in $mole^{-1}$ $l.^{+1}$ sec^{-1}, for reaction with		
	styrene	methyl methacrylate	vinyl acetate
polystyrene	176	338	3·2
polymethyl methacrylate	1600	734	36·7
polyvinyl acetate	$3\cdot7 \times 10^5$	$2\cdot47 \times 10^5$	$3\cdot70 \times 10^3$

It is now necessary to examine the statements that

(a) a polymer radical may be comparatively unreactive because it is stabilized by resonance,
(b) there is an inverse relationship between the reactivities of a monomer and the polymer radical derived from it.

Suppose that the propagation reaction in the ethylene polymerization

$$P.CH_2.CH_2\cdot + CH_2{:}CH_2 \longrightarrow P.CH_2.CH_2.CH_2.CH_2\cdot \qquad (36)$$

is taken as standard. In Fig. 4.4 the potential energy of the system is plotted against the separation of the reactants. Curve *AB* shows that the energy rises as the monomer approaches the radical; curve *CD* represents the increase in potential energy of the product radical with stretching of the bond joining the terminal monomer unit to the rest of the polymer radical. At the point of intersection of AB and CD, the potential energies of reactants and product are equal; the transition state is probably stabilized by resonance, and can be represented by a point X on the dotted line. The energy of activation for (36) and the heat of reaction are indicated in the diagram.

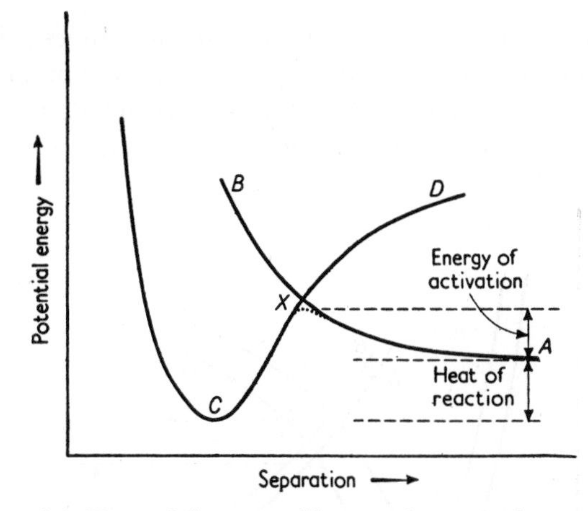

FIG. 4.4. Potential energy diagram for growth reaction.

Now consider a growth reaction involving the polyethylene radical and monomeric styrene to give a polystyrene radical

$$P.CH_2.CH_2\cdot + CH_2{:}CH.C_6H_5 \longrightarrow P.CH_2.CH_2.CH_2.CH(C_6H_5)\cdot \quad (37)$$

Because of resonance stabilization, the potential energy of the polystyrene radical is appreciably lower than that of the polyethylene radical; thus, in Fig. 4.5, curve CD refers to the product of (36), while the lower curve, $C'D'$, refers to the radical formed in (37).

Just as the phenyl group stabilizes the polystyrene radical with respect to the polyethylene radical, so to a lesser extent it also stabilizes the styrene molecule with respect to the ethylene molecule; curve $A'B'$ is, therefore, a little lower than AB. From Fig. 4.5:

$$(\text{heat of reaction } 37) = (\text{heat of reaction } 36) - x_m + x_r$$

where x_m = (potential energy of the ethylene molecule) − (potential energy of the styrene molecule), and x_r = the corresponding difference for the polymer radicals. If the pairs of curves (AB and $A'B'$) and (CD and $C'D'$) have very similar shapes, and if the effects due to stabilization of the transition state are the same for (36) and (37),

$$E_{36} - E_{37} = \alpha(x_r - x_m) \quad (38)$$

where α is positive but less than 1, its magnitude depending upon the slopes of AB and CD near the point of intersection (Evans, Gergely and Seaman, 1948).

If the pre-exponential factors for (36) and (37) are the same, (38) leads to

$$\frac{k_{37}}{k_{36}} = \frac{\exp(-E_{37}/RT)}{\exp(-E_{36}/RT)} = \exp\{\alpha(x_r - x_m)/RT\} \tag{39}$$

The velocity constants for reactions of all monomers with the polyethylene radical can be expressed in this way. If monomers are listed in order of reactivity towards this radical, the arrangement is the same as if they are placed in order of the value for $(x_r - x_m)$.

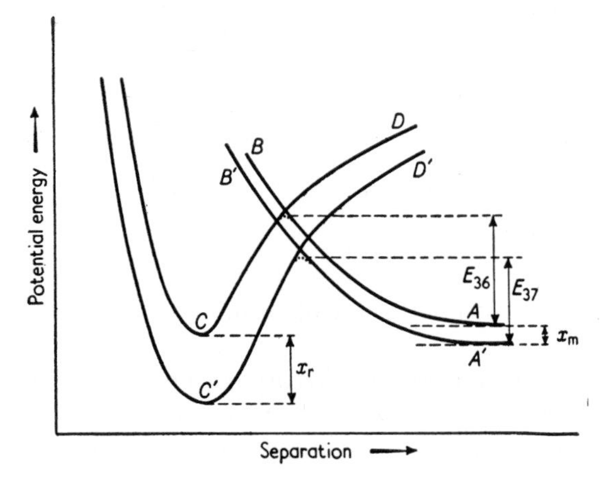

FIG. 4.5. Potential energy diagrams for reactions of a polymer radical with two monomers.

If the polyethylene radical is replaced by some other polymer radical as the reference radical, (39) is modified to

$$\frac{k_{37}}{k_{\text{ref.}}} = \frac{\exp\{\alpha(x_r - x_m)/RT\}}{\exp\{\alpha(x'_r - x'_m)/RT\}} \tag{40}$$

where x'_r and x'_m refer respectively to the reference radical and its related monomer, and $k_{\text{ref.}}$ refers to the reaction of the reference radical with its own monomer. The arrangement of monomers in order of reactivity should be the same for all reference radicals. The treatment is based on the belief that the reactivity of a polymer radical is governed solely by the nature of the monomer unit last added.

For comparison of the reactivities of polymer radicals towards reference monomers, the growth reaction in the ethylene polymerization (36) can again be taken as standard. Suppose that the polyethylene radical is replaced by a more stable polymer radical such as the polystyrene radical

$$P.CH_2.CH(C_6H_5)\cdot + CH_2{:}CH_2 \longrightarrow P.CH_2.CH(C_6H_5).CH_2.CH_2\cdot \quad (41)$$

The products in (36) and (41) can be regarded as the same, and so curve CD in Fig. 4.6 applies to both; $A'B'$ refers to the reactants in (41), and AB to those in (36), and so

$$E_{41} - E_{36} = x_r - x_t$$

If $(x_r - x_t)$ is expressed as γx_r, where γ is positive but less than 1, and if differences between k_{36} and k_{41} are due solely to differences between the corresponding activation energies,

$$k_{41}/k_{36} = \exp(-\gamma x_r/RT) \quad (42)$$

Just as in the comparison of the reactivities of monomers towards reference radicals, it is possible to arrange polymer radicals in order of

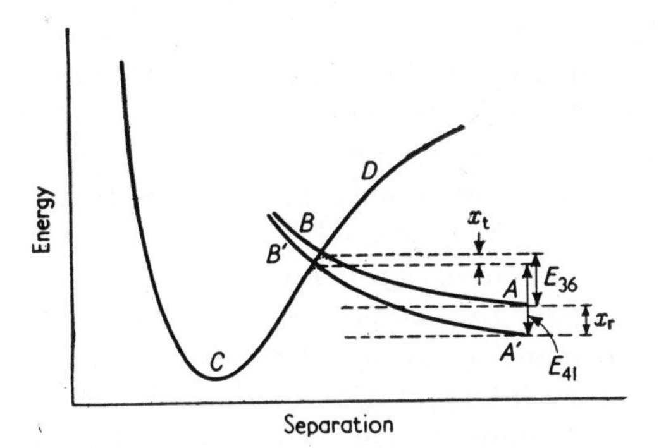

FIG. 4.6. Potential energy diagrams for reactions of a monomer with two polymer radicals.

reactivity towards a reference monomer. According to this treatment, the position of a radical on this scale is governed only by the value of x_r, and so it should be independent of the nature of the monomer. Inspection of (39) and (42) shows that if a polymer radical is highly stabilized (i.e. x_r large and positive), it is less reactive than the polyethylene radical while the monomer is more reactive than ethylene.

This treatment is based on a number of assumptions, including that differences between the rates of the various growth reactions are due to differences between the activation energies; this particular assumption could be tested experimentally, but unfortunately there is a scarcity of suitable data. If it were justified, all monomer reactivity ratios would tend towards unity at high temperatures; there are indications (Alfrey,

Bohrer and Mark, 1952) that this is so for some pairs of monomers. Some of the cases where it is obviously untrue involve 1,2-disubstituted ethylenes; other evidence indicates that steric factors are of great significance in polymerizations involving these monomers (see Section B.*5* of this Chapter).

4. Polar Effects in Growth Reactions

Serious exceptions to the prediction that the relative reactivities of polymer radicals and of monomer molecules should be independent of the nature of the reference monomer or radical, are associated with co-polymerizations in which both r_a and r_b are less than unity and $r_a r_b$ is very small. In these systems, cross-propagations are strongly favoured, and the monomer units tend to be arranged alternately in the resulting co-polymer molecules.

Mayo and Walling (1950) pointed out that alternation in binary co-polymerizations arises when one of the monomers contains an electron-attracting and the other an electron-repelling group, an example of such a pair being styrene and acrylonitrile. Alternation can reasonably be attributed to some sort of polar effect; it might be a dipolar attraction between radical and monomer, in which case a dependence of rate upon dielectric constant of the medium could be expected, but this has not been observed. It might be supposed that there is mutual polarization and, therefore, attraction between radical and monomer when their separation is very small; this would overcome the objection based upon the insensitivity of monomer reactivity ratios to environment. A third and more satisfactory interpretation of the polar effect supposes that polar structures contribute to the transition state, so that it can be represented as

$$\left[\begin{pmatrix}\text{polymer}\\ \text{radical}\end{pmatrix}^{\delta+}(\text{monomer})^{\delta-}\right] \quad \text{or} \quad \left[\begin{pmatrix}\text{polymer}\\ \text{radical}\end{pmatrix}^{\delta-}(\text{monomer})^{\delta+}\right]$$

depending upon the natures of the substituents in the monomer, and the terminal monomer unit in the radical. The existence of resonating structures would markedly reduce the energy of the transition state, and, therefore, also the activation energy for the reaction. It was assumed in Section B.*3* that any stabilization of the transition state would be the same for all growth reactions; if this essential assumption is invalid for certain cases, there must be deviations from the general pattern of behaviour.

Alfrey and Price (1947) developed the "Q and e" treatment for co-polymerizations to take account of both polar effects and the general reactivities of radicals and monomers. It is based on the assumptions that (*a*) alternation results from electrostatic interaction of charges on the radical and monomer,

(*b*) these charges are the same for a monomer and its related polymer radical.

Wall (1947) proposed a similar treatment in which assumption (*b*) is not made, but this is achieved only by introducing further parameters and extensive tests have not been made. The "Q and e" treatment must really be regarded as empirical although strikingly successful.

The velocity constant for cross-propagation is expressed in the form

$$k_{ab} = P_a Q_b \exp(-e_a e_b) \tag{43}$$

where P_a and Q_b are related to the reactivities of the (radical)$_a$ and (monomer)$_b$ respectively, e_a is a measure of the charge on (monomer)$_a$ or a

TABLE 4.9

Q AND e VALUES FOR MONOMERS AND POLYMER RADICALS

Monomer	Q	e	Reference monomer
styrene	1·0	−0·8	(standard)
methyl methacrylate	0·74	+0·4	styrene
acrylonitrile	0·44	+1·2	styrene
vinyl chloride	0·02	+0·2	styrene
	0·07	+0·4	methyl methacrylate
vinyl acetate	0·01	−0·5	vinyl chloride
	0·03	−0·3	methyl acrylate
	0·02	−0·1	vinylidene chloride

radical with this monomer unit at its reactive end, and e_b is defined similarly for (monomer)$_b$.

Thus

$$r_a = \frac{k_{aa}}{k_{ab}} = \frac{P_a Q_a \exp(-e_a^2)}{P_a Q_b \exp(-e_a e_b)} = \frac{Q_a}{Q_b} \exp\{-e_a(e_a - e_b)\} \tag{44}$$

and similarly

$$r_b = \frac{Q_b}{Q_a} \exp\{-e_b(e_b - e_a)\}$$

Price (1948) has taken ($Q = 1 \cdot 0$) and ($e = -0 \cdot 8$) for styrene as standards; Q and e values for other monomers may be calculated from monomer reactivity ratios. Quite apart from any inadequacies inherent in the treatment, errors in monomer reactivity ratios can lead to considerable uncertainties in the values of Q and e; typical values are shown in Table 4.9.

The "Q and e" treatment can be extended to systems of three or four components (Fordyce, Chapin and Ham, 1948). One of the chief uses of this treatment is for prediction concerning co-polymerizations not studied directly; uncertainties in the Q and e values lead to errors in the monomer reactivity ratios similar to those expected from direct determinations.

If e_a and e_b have the same sign and are about equal in magnitude, the exponential terms in (44) are close to unity, and r_a and r_b are governed by the Q values for the monomers. If, however, e_a and e_b have opposite signs, the exponential terms may be small so that r_a and r_b are small, and alternation in the co-polymer is very pronounced.

Another approach to the problem of polar effects in propagation reactions is due to Bamford, Jenkins and Johnston (1959*b*). The radical

TABLE 4.10

α AND β VALUES FOR MONOMERS

Monomer	α	β	Monomer	α	β
acrylonitrile	−3·1	5·3	methyl methacrylate	−1·4	4·9
methyl acrylate	−3·0	5·2	vinyl acetate	0	3·0
methacrylonitrile	−2·4	5·4	styrene	0	4·85
vinyl chloride	−1·4	3·65			

displacement reaction between a polymer radical and toluene (velocity constant, k_T) is taken as a process in which polar effects are likely to be of little importance. This reaction is chosen as the standard for the general reactivity of the polymer radical; from the experimental point of view, it is obviously much more convenient than the reaction between the polymer radical and ethylene discussed already. The velocity constant for the reaction of the polymer radical with any other substrate, whether a monomer or a transfer agent, can be expressed as

$$\log k = \log k_T + \alpha\sigma + \beta \quad (45)$$

α and β being constants for a given substrate, and σ being the algebraic sum of the Hammett substituent constants for the substituents in the terminal monomer unit of the radical. The values of α and β for various monomers are listed in Table 4.10.

In this treatment, β is a measure of the general reactivity of the monomer. The term $\alpha\sigma$ shows the extent to which polar effects contribute to the velocity constant for a reaction involving the polymer radical; its magnitude depends upon properties of both the radical and the substrate.

The treatment is considered again in connection with transfer reactions. Application to the competition between growth reactions in a copolymerization

$$P_a\cdot + M_a \longrightarrow P_a\cdot$$
$$P_a\cdot + M_b \longrightarrow P_b\cdot$$

gives

$$\log r_a = \log(k_{aa}/k_{ab}) = (\alpha_a - \alpha_b)\sigma_a + (\beta_a - \beta_b) \quad (46)$$

where α_a and β_a refer to M_a, α_b and β_b to M_b, and σ_a to $P_a\cdot$,

and
$$\log r_b = (\alpha_b - \alpha_a)\sigma_b + (\beta_b - \beta_a)$$

TABLE 4.11

σ VALUES FOR RADICALS

Radical	σ
benzoyloxy	approx. 1·0
polyacrylonitrile	0·66
polymethacrylonitrile	0·49
polymethyl acrylate	0·45
polyvinyl acetate	0·31
polymethyl methacrylate	0·28
polystyrene	−0·01

The term $(\beta_a - \beta_b)$ depends upon the general reactivities of the monomers, and the terms involving α and σ represent the influence of polar effects. When (46) is rearranged in the form

$$r_a = \frac{\exp(2\cdot303\beta_a)}{\exp(2\cdot303\beta_b)} \exp\{2\cdot303\sigma_a(\alpha_a - \alpha_b)\}$$

a similarity with (44), derived from the "Q and e" treatment, can be recognized.

This treatment can be applied to reactions of certain small radicals with monomers. Using the data in Table 3.1 for the relative reactivities of the benzoyloxy radical with various monomers for which α and β are known, it is calculated that for

$$C_6H_5.CO.O\cdot + C_6H_5.CH_3 \longrightarrow C_6H_5.COOH + C_6H_5.CH_2\cdot \quad (47)$$
$$C_6H_5CO.O\cdot \longrightarrow C_6H_5\cdot + CO_2 \quad (48)$$

k_{47}/k_{48} is approximately 2×10^{-4} mole^{-1} l.$^{+1}$ at 60°C. The value derived for σ for the benzoyloxy radical is shown in Table 4.11 together with those for various polymer radicals; the high value for the benzoyloxy radical shows that polar effects are very important in its reactions.

5. *Steric Effects*

Steric effects may be significant in growth reactions, particularly those involving 1,2-disubstituted ethylenes. These compounds show little tendency to form homo-polymers, and in co-polymers sequences of two or more disubstituted monomer units are very rare. Long-chain compounds having the repeating unit ·CHX.CHY· are stable unless the substituents are very bulky; the failure of the corresponding monomer to polymerize must therefore result from kinetic and not thermodynamic factors. There is probably steric interference between the carbon atoms C′ and C″ in the transition state for the reaction

$$\text{P.C'HX.CHY}\cdot + \text{C''HX:CHY} \longrightarrow \text{P.C'HX.CHY.C''HX.CHY}\cdot \qquad (49)$$

It is instructive to contrast the behaviours of 1,2- and 1,1-disubstituted ethylenes in polymerization. Heats of polymerization for monomers of the latter type, such as methyl methacrylate, are appreciably lower than those for monosubstituted ethylenes, and the bulky substituents make it difficult to construct molecular models. These facts suggest that in the polymer chain

$$\diagdown CH_2 \diagup CXY \diagdown CH_2 \diagup CXY \diagdown CH_2 \diagup CXY \diagdown$$

there is appreciable interference between the substituents on the alternate carbon atoms. For head-to-tail addition, the carbon atoms corresponding to C′ and C″ in (49) carry only hydrogen atoms so that steric interference in the transition state can usually be neglected.

An important extension to the treatment of steric effects in co-polymerizations concerns the relative reactivities of *cis* and *trans* isomers of 1,2-disubstituted ethylenes (Lewis and Mayo, 1948). In some cases, there are profound differences; for example, at 60°C *trans* 1,2-dichlorethylene is about six times more reactive than the *cis* isomer towards the polyvinyl acetate and polystyrene radicals (Doak, 1948), suggesting that for the addition reactions

$$E_{cis} - E_{trans} = \text{approx. 1 kcal/mole}$$

Referring to Fig. 4.7, the difference between the activation energies is given by

$$E_a - E_b = (t_a - t_b) - (i_a - i_b)$$

where t_a and t_b are the heat contents for the transition states of the two reactions, and i_a and i_b are the heat contents for the corresponding initial states. The product radicals from the *cis* and *trans* isomers are identical because of rotation about the carbon–carbon bond, and so normally it would be expected that the transition state would be intermediate between the initial and final states, and that

$$(t_a - t_b) < (i_a - i_b)$$

Under these conditions, E_b is greater than E_a and the less stable isomer is the more reactive; this is the situation for the 1,2-dichlorethylenes, for which the *trans* isomer is the less stable. If, however,

$$(t_a - t_b) > (i_a - i_b)$$

then E_b is less than E_a and the more stable isomer is the more reactive. Lewis and Mayo (1948) found cases for which this result applies; for

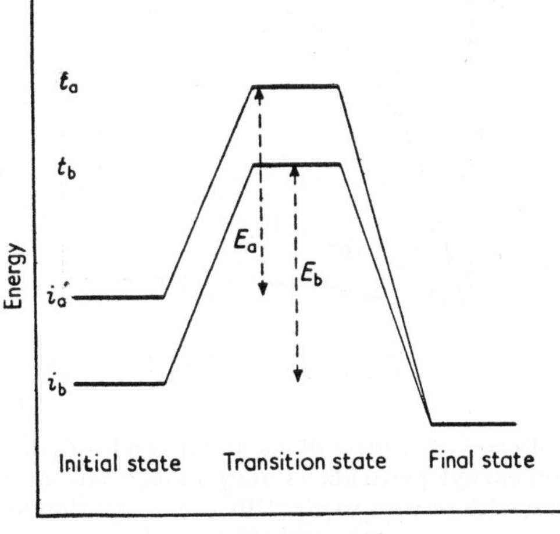

FIG. 4.7. Energy diagrams referring to the reactions involving a radical and the *cis* and *trans* isomers of a 1,2-disubstituted ethylene.

example, diethyl fumarate is considerably more reactive in radical reactions, and thermodynamically more stable than the corresponding maleate. In the transition state for reactions involving the fumaric ester, there is little interference between the two ester groups and a configuration permitting resonance is possible; the consequent stabilization lowers the activation energy for the reaction of this isomer. This effect is likely to be significant for any 1,2-disubstituted olefin in which the substituents are rather large, and the greater reactivity of the more stable *trans* isomer has been observed also in co-polymerizations involving stilbene.

Similar conclusions have been reached from examination of the reactivities of the isomers of stilbene towards the benzoyloxy radical

(Bevington and Brooks, 1958). The more stable *trans* isomer is the more effective in suppressing the formation of carbon dioxide from di-benzoyl peroxide (see Fig. 4.8). This technique would be very suitable for comparing the reactivities of the isomers over a range of temperatures, in order to test the view that the difference is due mainly to the activation energy term. It offers obvious advantages over the co-polymerization method, which is elaborate and depends upon accurate determinations of monomer reactivity ratios.

The reactivities of *cis* and *trans* isomers have also been compared in the co-polymerizations of the 2-butenes with sulphur dioxide (Dainton,

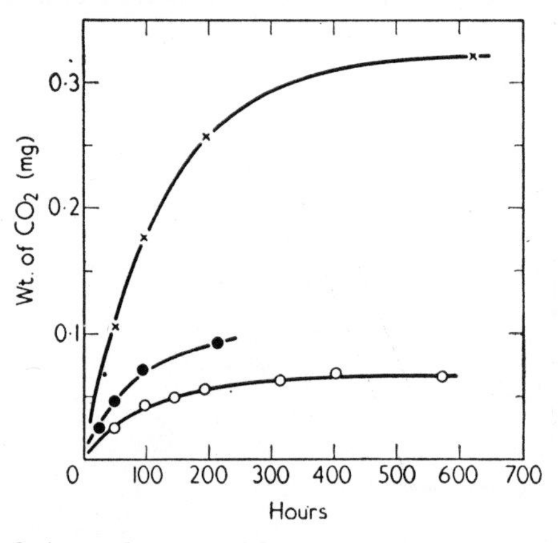

FIG. 4.8. Effects of *cis*- and *trans*-stilbene upon yields of carbon dioxide during decomposition of dibenzoyl peroxide (1 mg) in benzene at 60°C. x benzene; ● solution also 0·28 M with respect to *cis*-stilbene; ○ solution 0·27 M with respect to *trans*-stilbene.

Diaper, Ivin and Sheard, 1957; Ivin, Keith and Mackle, 1959). It was possible to obtain satisfactory correlation between the activation energies for the propagation and depropagation reactions, and the difference between the heat contents of the isomers.

6. *Non-ideal Systems*

In a few co-polymerizations, the monomer reactivity ratios apparently depend upon the composition of the feed. This effect can be accounted for by supposing that penultimate monomer units can influence the reactivities of co-polymer radicals. It may be necessary (Merz, Alfrey and Goldfinger, 1946) to consider four types of polymer radical in a binary co-polymerization, viz. $P.M_a.M_a\cdot$, $P.M_b.M_a\cdot$, $P.M_a.M_b\cdot$ and $P.M_b.M_b\cdot$, and eight growth reactions, given the velocity constants k_{aaa}, k_{aab}, k_{baa},

k_{bab}, k_{aba}, k_{abb}, k_{bba} and k_{bbb}. Four monomer reactivity ratios could then be defined:

$$r_a = \frac{k_{aaa}}{k_{aab}} \qquad r_a' = \frac{k_{baa}}{k_{bab}}$$

$$r_b = \frac{k_{bbb}}{k_{bba}} \qquad r_b' = \frac{k_{abb}}{k_{aba}}$$

In the absence of effects due to non-terminal monomer units, r_a' is equal to r_a, and r_b' to r_b. If penultimate groups can influence the reactivities of polymer radicals, units further from the reactive end also may exert effects; Ham (1960) has published a thorough kinetic analysis and applied it to some particular systems in which influences of non-terminal units may be quite marked.

A penultimate group effect has been recognized in the co-polymerization of styrene and sulphur dioxide; not only does sulphur dioxide fail to add to a radical with a terminal SO_2 unit, but it also will not react with a polymer radical in which SO_2 is the penultimate unit (Barb, 1953*a*; Walling, 1955). In the co-polymerization of fumaronitrile with styrene (Ham, 1954), the velocity constant for the cross-reaction

$$P.CH_2.CH(C_6H_5)\cdot + CH(CN){:}CH(CN) \longrightarrow P.CH_2.CH(C_6H_5).CH(CN).CH(CN)\cdot \quad (50)$$

is greatest when the concentration of the nitrile in the group P is low. In this system, there may even be small effects due to fumaronitrile units six or eight places away from the reactive end of a polymer radical (Ham, 1960). There are penultimate group effects during the co-polymerizations of styrene and maleic anhydride (Barb, 1953b) and styrene and oxygen (Mayo, Miller and Russell, 1958). The radical

$$P.CH_2.CH(C_6H_5).CH_2.CH(C_6H_5)\cdot$$

is about ten times as reactive as

$$P.O.O.CH_2.CH(C_6H_5)\cdot$$

towards oxygen.

Penultimate group effects have also been detected in a few co-polymerizations of monomers, both of which can polymerize separately; for example, in the system styrene–acrylonitrile, the reactivity towards acrylonitrile of a radical having a styrene terminal unit is damped as the acrylonitrile content of the co-polymer increases (Ham, 1954).

Another reason why the co-polymer composition equation (28) may break down is that some, or all, of the propagation steps may be reversible, i.e. depropagation may be significant. This situation is bound to arise for all co-polymerizations if the temperature is sufficiently high, but then, in most cases, complications are likely to arise from side-reactions involving pendant groups in the co-polymers. Co-polymerizations of sulphur dioxide and olefins have low ceiling temperatures and depolymerization may become significant at comparatively low temperatures. It is possible also that an effect might be found in co-polymerizations involving methacrylate esters at temperatures in the region of 130°C, where depropagation can be detected in the homo-polymerizations of these monomers.

The co-polymerization of sulphur dioxide with either of the isomers of 2-butene is accompanied by isomerization of the olefin (Bristow and Dainton, 1955); the rate of isomerization increases as the temperature approaches the ceiling temperature although the rate of polymerization decreases. The effect can be accounted for in terms of the propagation-depropagation equilibria

$$P.SO_2\cdot + \begin{matrix} H \\ CH_3 \end{matrix}\!\!>C{=}C<\!\!\begin{matrix} CH_3 \\ H \end{matrix} \rightleftharpoons P.SO_2.CH(CH_3).CH(CH_3)\cdot$$

$$\rightleftharpoons P.SO_2\cdot + \begin{matrix} H \\ CH_3 \end{matrix}\!\!>C{=}C<\!\!\begin{matrix} H \\ CH_3 \end{matrix} \qquad (51)$$

In most co-polymerizations involving sulphur dioxide, only 1:1 co-polymers are produced, either because the reaction is really the self-polymerization of a 1:1 complex of olefin and sulphur dioxide, or because it is a true co-polymerization in which both r_a and r_b are zero; in either case, occurrence of depolymerization cannot affect the composition of the product. In the co-polymerization of styrene with sulphur dioxide, however, products having a range of compositions are possible and deviations from (28) may be expected; they are quite distinct from the penultimate group effect already discussed. If [sulphur dioxide] in the feed is held constant while [styrene] is raised, the ratio $[SO_2]/[C_8H_8]$ in the co-polymer rises. Barb (1953*a*) and Walling (1955) supposed that the reactions (52) and (53), both of which involve the formation of C—S bonds, are reversible:

$$P.SO_2\cdot + CH_2{:}CH.C_6H_5 \longrightarrow P.SO_2.CH_2.CH(C_6H_5)\cdot \qquad (52)$$

$$P.CH_2.CH(C_6H_5)\cdot + SO_2 \longrightarrow P.CH_2.CH(C_6H_5).SO_2\cdot \qquad (53)$$

The product of (52) may dissociate or it may take up another molecule of styrene by a reaction which is irreversible under ordinary conditions:

$$P.SO_2.CH_2.CH(C_6H_5)\cdot + CH_2{:}CH.C_6H_5 \longrightarrow P.SO_2.CH_2.CH(C_6H_5).CH_2.CH(C_6H_5)\cdot \qquad (54)$$

(52) and (54) are favoured by high concentrations of styrene; once (54) has occurred, the sulphur dioxide is locked in the co-polymer.

The co-polymerization of ethylene and carbon monoxide is another system in which depropagation may be significant (Walling, 1955). Small acyl radicals, e.g. $CH_3.CO\cdot$ and $C_6H_5.CO\cdot$, readily lose carbon monoxide although such dissociations can be suppressed at high pressures (Walling and Savas, 1960); it is likely therefore that the growth reaction

$$P.CH_2.CH_2\cdot + CO \longrightarrow P.CH_2.CH_2.CO\cdot \qquad (55)$$

is reversible. Kinetic analysis based on this belief shows that it is possible to account for the composition of the co-polymer depending upon the absolute concentration of carbon monoxide, and not upon the ratio of the concentrations of the two monomers in the feed; further, a very reasonable value for the activation energy for the reverse of (55) can be derived. If ethylene is replaced by another monomer, the activation energy for

$$P.CH_2.CHX\cdot + CO \longrightarrow P.CH_2.CHX.CO\cdot \qquad (56)$$

is increased and that for the reverse process decreased, because of resonance stabilization of the radical $P.CH_2.CHX\cdot$; as a result, carbon monoxide will not co-polymerize with monomers other than ethylene. Barb (1953*c*) regarded the ethylene–carbon monoxide co-polymerization as one involving ethylene and a 1:1 complex of ethylene and carbon monoxide. The observed dependence of composition of co-polymer upon composition of the feed is readily explained, but no account is taken of the known instability of acyl radicals.

C. Dienes and Divinyl Monomers

Attention is now directed to monomers having more than one olefinic bond per molecule. A point of special interest is that, because of their polyfunctionality, these monomers can give rise to branched and cross-linked polymers; another feature is that the monomers in which the double bonds are conjugated, can give polymers containing more than one type of monomer unit.

In polymers of butadiene, there may be three types of monomer unit, viz.

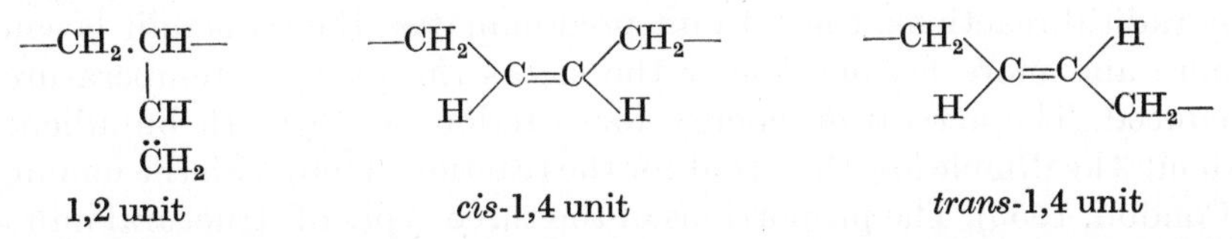

and spectroscopic and chemical methods have been developed to determine their relative numbers. The 1,2 unit possesses an asymmetric carbon atom so that both *d* and *l* forms may exist. In substituted butadienes, such as isoprene, yet another type of monomer unit is possible since the 1,2 and 3,4 units are different:

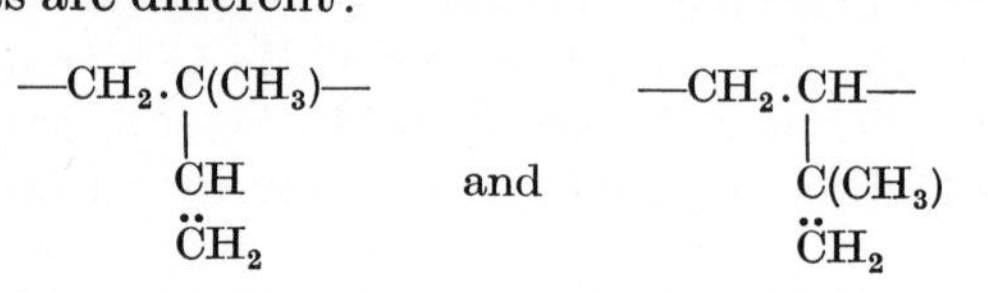

It is important to recognize just how the various types of monomer unit are produced during polymerization. Attack of a radical upon butadiene occurs at a CH_2 group giving (57), but this structure is related to others thus:

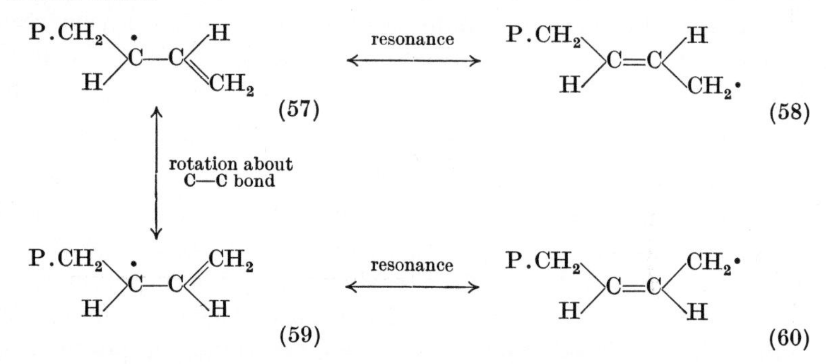

If the radical takes up another monomer molecule by reacting in the form (57) or (59), a 1,2 unit is produced in the polymer; if it reacts as (60), a *cis*-1,4 unit is formed, whereas a *trans*-1,4 unit results from reaction in the form (58). The structure of a particular butadiene unit in the polymer chain is not fixed when this unit enters the polymer but only when the next monomer is added. The various types of unit in the polymer result from the polymer radical being able to react in more than one way.

The proportions of the various types of unit in polybutadiene and similar polymers depend upon the mechanism of polymerization, i.e. radical or ionic; in radical systems, the proportions are not affected by the method of polymerization (bulk or emulsion), the nature of the initiator or the conversion, but they are sensitive to the temperature at which the polymerization is performed. In polybutadiene, prepared by radical reactions, the 1,4 unit predominates; the *trans* unit becomes more and more favoured over the *cis* as the reaction temperature is reduced. The activation energy for reaction of (58) with butadiene is about 3 kcal/mole less than that for the reaction of (60) with the monomer (Condon, 1953). The proportions of the three types of structural unit are governed by the relative rates of the reactions of (57), (58), (59) and (60) with monomer; these may well be affected by the precise nature of the monomer, so that the proportions in pure polybutadiene may be different from those in co-polymers containing butadiene. Fig. 4.9 shows that with rising styrene content in a co-polymer, the proportion of *trans*-1,4 butadiene units increases at the expense of the *cis*-1,4 and 1,2 units. Effects with co-monomers other than styrene are similar but differ in magnitude (Foster and Binder, 1953).

In polyisoprene prepared by radical processes, 1,2 and 3,4 units each account for about 5% of the total number of monomer units; these proportions vary little with temperature. The percentage of *cis*-1,4 units rises from close to zero for polymers prepared at 0°C, to about 25% for those made at 100°C (Richardson and Sacher, 1953).

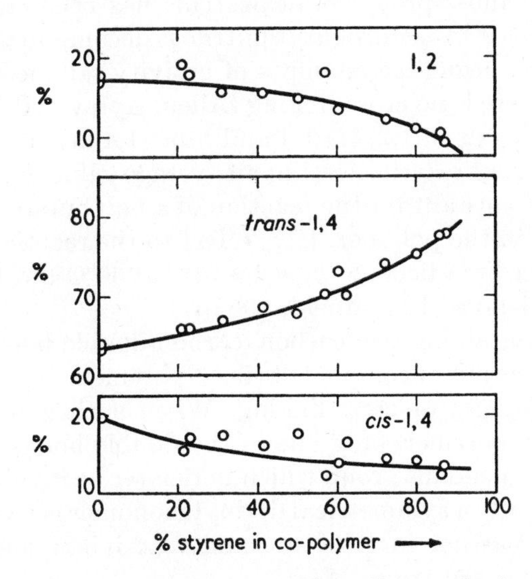

Fig. 4.9. Proportions of various types of butadiene units in co-polymers with styrene (Foster and Binder, 1953).

Co-polymerization studies reveal that butadiene is a reactive monomer with Q equal to $1 \cdot 33$; this is due to resonance stabilization of the polymer radical, involving structures (57)–(60).

Both the 1,2 and 1,4 units in polybutadiene contain double bonds which are susceptible to radical attack to give

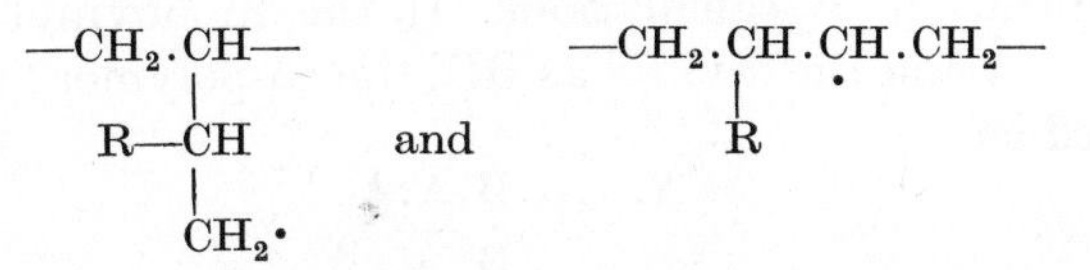

These products are not appreciably stabilized by resonance, so that the double bonds in diene polymers should be less reactive than the double

bonds in the monomer; this has been confirmed by studies of graft co-polymerizations involving polyisoprene as stock. If methyl methacrylate is polymerized in the presence of a polyisoprene using dibenzoyl peroxide as initiator, side-chains are attached to the polymer; if, however, azo*iso*butyronitrile is used as initiator, no grafting occurs (Allen and Merrett, 1956). Evidently, the grafting does not result from reaction of polymethyl methacrylate radicals with double bonds in the polyisoprene, otherwise the nature of the initiator should have no effect on the reaction. It must be concluded that although the double bonds are reactive enough to capture phenyl and benzoyloxy radicals, they do not react with 2-cyano-2-propyl radicals. This has been confirmed by the observation that if ^{14}C-azo*iso*butyronitrile is decomposed in the presence of polyisoprene, negligible amounts of activity are incorporated in the polymer and there is no cross-linking (Allen, Ayrey and Moore, 1959).

The double bonds in 1,2, *cis*-1,4 and *trans*-1,4 units in polybutadiene must have different reactivities, but an average value, k_x, can be assigned to the velocity constant for the reaction of a polybutadiene radical with a double bond in the polymer. If k_p refers to the reaction of the radical with monomeric butadiene, k_x/k_p is 2×10^{-4} at 60°C and 1×10^{-4} at 40°C (Morton, Salatiello and Landfield, 1952).

Monomers containing two carbon–carbon double bonds per molecule have been used as minor components in co-polymers to produce materials with known degrees of cross-linking. When such a monomer is first incorporated in polymer, only one of the double bonds reacts, and the other is left as a pendant group which may later be involved in a growth reaction. Generally a symmetrical divinyl monomer is chosen resembling its monovinyl partner in structure and reactivity; some of the pairs which have been used are:

divinyl benzene, $CH_2:CH.C_6H_4.CH:CH_2$, with styrene;

ethylene glycol dimethacrylate, $CH_2:C(CH_3).CO.O.CH_2.CH_2.O.CO.C(CH_3):CH_2$, with methyl methacrylate;

divinyl adipate, $CH_2:CH.O.CO.[CH_2]_4.CO.O.CH:CH_2$, with vinyl acetate.

By this means, the chemical compositions of the co-polymer and the homo-polymer of the monovinyl compound are very similar.

Another reason for choosing co-monomers which resemble one another, is for simplification of calculations. If the monovinyl compound is represented as A and the divinyl as BB, the co-polymer first formed can be formulated as

$$\begin{array}{c} \text{—A.A.A.B.A.A—} \\ \dot{\text{B}} \end{array}$$

If the reactivities of the double bonds in A, BB and the pendant B groups are the same, it is quite easy to calculate the probability that a pendant group engages in reaction (Alfrey, Bohrer and Mark, 1952). These authors have also considered more general cases in which the double bonds of the vinyl and divinyl monomers do not have equal reactivities. Co-polymerizations involving dienes form a special case; the two double bonds in the monomeric diene are of equal reactivity, but once one of them has reacted the residual double bond has a much lower reactivity.

It is instructive to consider in more detail the three symmetrical divinyl monomers mentioned already. Monomeric divinyl benzene is even more reactive than monomeric styrene because it gives rise to a polymer radical considerably stabilized by resonance; for the same reason, a radical with divinyl benzene as its terminal unit is likely to be rather unreactive. Such a radical undoubtedly reacts mainly in the form

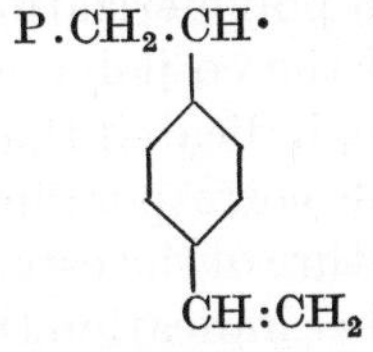

$P.CH_2.CH\cdot$

$CH:CH_2$

Subsequent reaction of the pendant vinyl group with a polymer radical gives a product represented as

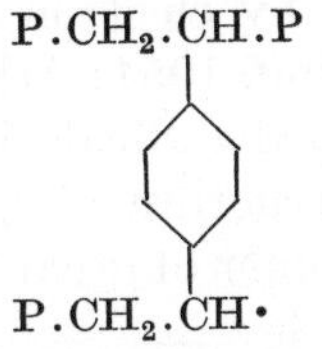

$P.CH_2.CH.P$

$P.CH_2.CH\cdot$

This radical must be resonance stabilized to about the same extent as the polystyrene radical. A double bond in polydivinyl benzene is therefore less reactive than a double bond in the original monomer; the effect resembles that found for the dienes.

In ethylene glycol dimethacrylate and divinyl adipate, the double bonds are not conjugated and, therefore, do not influence one another. These monomers, therefore, with methyl methacrylate and vinyl acetate respectively, form systems quite suitable for studies of cross-linking; Walling (1945) used them for tests of the theory of gelation. Except at low concentrations of the divinyl monomer, cross-linking was less frequent than expected. This may have been due to the occurrence of the "intramolecular" reaction of a pendant carbon–carbon double bond with the free radical at the end of the polymer chain to which it was itself attached. Polymer molecules are certainly not fully extended in solution, and coiling must occur for polymer radicals also; this would tend to favour intramolecular cross-linking over intermolecular cross-linking, because the distribution of pendant double bonds in the reaction mixture is not random with respect to the polymer radicals. It is significant that Haward (1950) showed that termination in a diradical polymerization is quite likely to occur by the interaction of the radicals at the two ends of a single chain.

In Section A.*3* it was shown that the growth reaction in an ordinary polymerization becomes diffusion-controlled only at fairly high viscosities since it involves a small monomer molecule which diffuses freely; if the double bond, however, is attached to a large polymer molecule, the growth reaction occurs between two large particles, and it is much more susceptible to diffusion-control. This effect also would lead to a favouring of intra-molecular cross-linking.

Comparisons of the homo-polymerizations of methyl, *n*.butyl and propyl methacrylates and of the co-polymerizations of the first two of these monomers with styrene, indicated that the reactivities of an ester of methacrylic acid, and the corresponding polymer radical, are not strongly dependent on the nature of the ester group (Burnett, Evans and Melville, 1953). This is a further indication that the co-polymerization of ethylene glycol dimethacrylate with methyl methacrylate is one very suitable for study of cross-linking. Gels formed during this co-polymerization have been examined by electron spin resonance spectroscopy (Fraenkel, Hirshon and Walling, 1954; Atherton, Melville and Whiffen, 1959). The presence of trapped radicals has been confirmed, and the work gives evidence for the structure of the growing radicals.

A comparatively new monomer of great interest is *p*.xylylene:

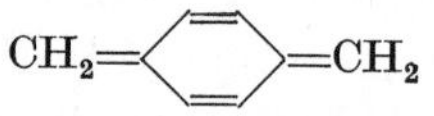

$$CH_2{=}C_6H_4{=}CH_2$$

This material has conjugated carbon–carbon double bonds, and it is structurally similar to *p*.benzoquinone; its chemistry has been reviewed by Errede and Szwarc (1958). It is the simplest member of a group of related compounds. Although stable in the gas phase and at very low temperatures, under other conditions *p*.xylylene polymerizes very readily to a product of high molecular weight; the physical and chemical properties of the polymer (Brown and Farthing, 1953) are consistent with a repeating unit

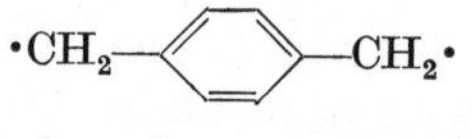

$$\cdot CH_2{-}C_6H_4{-}CH_2\cdot$$

It seems to have been taken for granted that the polymerization of *p*.xylylene is a diradical process, initiated by a spontaneous dimerization:

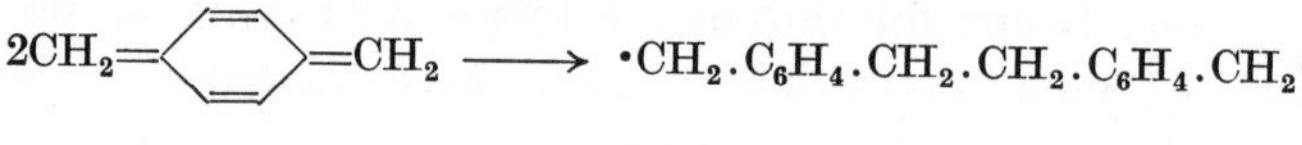

$$2CH_2{=}C_6H_4{=}CH_2 \longrightarrow \cdot CH_2.C_6H_4.CH_2.CH_2.C_6H_4.CH_2\cdot$$

In previous sections, it has been pointed out that attempts to polymerize vinyl monomers by diradical mechanisms have failed; usually this has been attributed to termination by cyclization after the addition of only a few monomer units. The presence of benzene rings in the main chain of the *p*.xylylene polymer may make ring closure less probable; further, polymerization gives an insoluble product of high crystallinity in which there is likely to be only very restricted motion of the segments of the polymer chains. As yet, there are no firm conclusions concerning the mechanism of termination in the polymerization, and no data concerning chain transfer. It appears that *p*.xylylene does not co-polymerize with monomers such as styrene, although it readily forms an alternating co-polymer with oxygen (Errede and Hopwood, 1957), and co-polymers with substances of the same class as itself. Solutions, which can be preserved at $-80°C$, cannot be stabilized at higher temperatures by those substances which prevent radical polymerization of vinyl monomers. There is, in fact, no compelling evidence that the polymerization of *p*.xylylene really is a radical process; the reaction clearly merits much more attention, and the attractive physical and chemical properties of the polymer may encourage this.

Chapter 5

Transfer Reactions

A. General Features

Transfer reactions are two-stage processes, radical-displacement (1) being followed by re-initiation (2):

$$P\cdot + A.B \longrightarrow P.A + B\cdot \tag{1}$$
$$B\cdot + M \longrightarrow B.M\cdot \tag{2}$$

Usually A is a hydrogen atom, so that the first stage is hydrogen abstraction, but it may be a halogen atom or in certain cases a group of atoms. When considering the actual course of radical-displacement, it must be recognized that the reaction may involve any univalent atom or group of atoms in the molecule of the transfer agent; for reactive transfer agents, however, one reaction path is usually much favoured.

If the probability of the radical B· being captured by monomer is the same as that of a growing polymer radical acquiring another monomer unit, the total rate of polymerization is hardly affected, and the only obvious effect of the transfer process is a reduction in the average molecular weight of the resulting polymer. The substance AB may be any component of the reaction mixture, monomer, diluent, initiator or dead polymer; in some cases, small amounts of particularly reactive substances may be added to the reaction mixture to control the molecular weight of the resulting polymer.

If (2) is slow, the stationary concentration of B· builds up and the reactions

$$2B\cdot \longrightarrow \text{products of combination or disproportionation} \tag{3}$$
$$P\cdot + B\cdot \longrightarrow P.B \tag{4}$$
$$P\cdot + B\cdot \longrightarrow \text{products of disproportionation} \tag{5}$$

become significant. In this case, the transfer agent acts also as a retarder, reducing the kinetic chain length and the overall rate of polymerization as well as the average molecular weight of the polymer; this type of reaction is known as *degradative transfer*.

The ratio k_1/k_p is usually found by study of the number average molecular weights of polymers; the ratio is referred to as the transfer constant for the polymer radical with the particular substance. For transfer to monomer, k_1 is usually written as k_f. The number-average degree of polymerization of a sample of polymer is defined by

$$\bar{P} = \frac{\text{number of monomer units}}{\frac{1}{2}(\text{number of chain ends})}$$

For polymers of reasonably high molecular weight, the number of monomer units is controlled by the propagation process. The number of chain ends depends upon termination and transfer reactions. If transfer can be neglected,

rate of formation of chain ends =
f. (rate of initiation of polymerization)

where $f = 1$ if termination is entirely by combination, and $f = 2$ if termination is by disproportionation. Each transfer reaction increases the number of chain ends by 2, provided that all B• radicals are consumed in either (2), (4) or (5). In the steady state,

$$\bar{P} = \frac{k_p[\text{P}\cdot][\text{M}]}{\frac{1}{2}\{f.R_i + 2\sum(\text{rates of transfer reactions})\}}$$

$$= \frac{k_p[\text{P}\cdot][\text{M}]}{\frac{1}{2}f.R_i + k_f[\text{P}\cdot][\text{M}] + k_1[\text{P}\cdot][\text{AB}] + k_1'[\text{P}\cdot][\text{I}]}$$

and

$$\frac{1}{\bar{P}} = \frac{k_f}{k_p} + \frac{k_1[\text{AB}]}{k_p[\text{M}]} + \frac{k_1'[\text{I}]}{k_p[\text{M}]} + \frac{\frac{1}{2}f.R_i}{k_p[\text{P}\cdot][\text{M}]} \quad (6)$$

where [AB] = concentration of transfer agent,
[I] = concentration of initiator, and k_1' refers to the radical-displacement in which it is involved.

The stationary concentration of growing polymer radicals is given by

$$R_p = k_p[\text{P}\cdot][\text{M}]$$

Normally, the rate of initiation is proportional to $[\text{initiator}]^1$ and the rate of polymerization to $[\text{initiator}]^{1/2}$, so (6) can be converted to (7):

$$\frac{1}{\bar{P}} = \frac{k_f}{k_p} + \frac{k_1[\text{AB}]}{k_p[\text{M}]} + \frac{k_1'[\text{I}]}{k_p[\text{M}]} + \text{constant}.[\text{I}]^{1/2} \quad (7)$$

This equation forms the basis for determining transfer constants for monomers, additives and initiator; the procedures have been very adequately discussed by Bamford, Barb, Jenkins and Onyon (1958). When measuring the number average molecular weights, caution must be exercised in the use of indirect methods such as viscometry; transfer reactions may modify the distribution of molecular weights in the polymer as well as altering the average value.

According to the scheme presented, the transfer agent becomes included in the polymer; if

$$\text{rate of (1)} = \{\text{rate of (2)}\} + \{\text{rate of (4)}\} \tag{8}$$

each transfer step causes one molecule of the transfer agent to become incorporated in polymer. The process can be likened to co-polymerization, although the two parts of the transfer agent become included in different polymer molecules. If

(*a*) the kinetic chains are long so that consumption of monomer in the initiation step is negligible,
(*b*) primary radical transfer is neglected,
(*c*) [monomer] and [transfer agent] do not change appreciably during the reaction,

then

$$\frac{\text{no. of monomer units in polymer}}{\text{no. of molecules of transfer agent in polymer}} = \frac{k_p[\text{P}\cdot][\text{M}] + k_2[\text{B}\cdot][\text{M}]}{k_1[\text{P}\cdot][\text{AB}]}$$

When transfer occurs infrequently, the second term in the numerator can be neglected and the right-hand side of the equation simplifies to $k_p[\text{M}]/k_1[\text{AB}]$. Analysis of the polymer could lead, therefore, to a value for the transfer constant.

In certain cases, the analytical method can lead to quite erroneous values for transfer constants, for example, giving high values if the agent can also be incorporated in polymer by reactions other than transfer; this limitation applies to benzene which may co-polymerize with certain monomers (Stockmayer and Peebles, 1953). The first stage in the reaction is

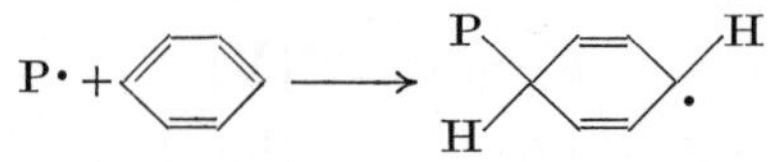

and the product can react with monomer. When bromobenzene is used with styrene at 156°C, it is thought that a similar adduct is formed (Mayo, 1953) but when it reacts with another molecule of monomeric styrene the reaction

$$\text{P.CH}_2\text{.CH(C}_6\text{H}_5\text{)(H)C}_6\text{H}_5\text{Br}\cdot + \text{CH}_2\text{:CH.C}_6\text{H}_5$$

$$\longrightarrow \text{P.CH:CH.C}_6\text{H}_5 + \text{C}_6\text{H}_5\text{Br} + \text{CH}_3\text{.CH(C}_6\text{H}_5)\cdot$$

is believed to occur instead of addition. This reaction scheme explains why the analytical method gives a much lower value for k_1/k_p than the molecular weight method in this system.

For determination of transfer constants by analysis of polymers, sensitive methods are essential; labelled transfer agents might be used, but care must be exercised in interpreting the results. If the labile hydrogen atom in a transfer agent is labelled with a heavy isotope, the rate of reaction is decreased because of an isotope effect. If tritium is used for labelling, the fraction of the molecules actually containing the heavy isotope would normally be small, and there would be no significant difference between the molecular weights of polymers prepared in the presence of labelled and unlabelled transfer agents at the same concentration. The molecule $B.{}^{1}H$ would, however, react more readily than $B.{}^{3}H$, and measurement of the rate of incorporation of tritium in the polymer would lead to a low value for the transfer constant. If the proportion of molecules actually containing the heavy hydrogen is large, as would normally be the case if deuterium is used as a label for hydrogen, there would be a significant difference between the molecular weights of polymers prepared in the presence of labelled and unlabelled transfer agents at the same molar concentration. These effects have been found (Bevington and Troth, 1960) when using deuterated and tritiated samples of triphenylmethane as transfer agents.

If (8) is satisfied, a transfer agent labelled in the group B can be used in the analytical method for determining transfer constants. This principle has been used in studies of mercaptans as transfer agents using materials labelled with sulphur-35. If, however, an appreciable proportion of the B· radicals engage in (3) or (5), the transfer constant will be underestimated since the rate of radical-displacement is greater than the rate of entry of B· radicals into polymer.

For systems containing more than one transfer agent, it is necessary to add extra terms to (7). Nandi and Palit (1960) confirmed that the effect of a mixture of transfer agents upon the degree of polymerization can be predicted from the separate transfer constants. The use of mixed transfer agents would be valuable for study of the effects of substances which would, by themselves, cause precipitation of the polymer and resulting kinetic complications; thus, transfer constants for water could be determined from studies of the molecular weights of polymers prepared in solvents which can dissolve appreciable quantities of water without causing precipitation of polymer. Jenkins and Johnston (1959) showed, however, that in the polymerization of acrylonitrile in dimethylformamide, the addition of water reduces the reactivity of the solvent as a transfer agent; apparently a hydrate of dimethylformamide is formed and is comparatively inert towards the growing radicals. Isotopically labelled transfer agents might be valuable for study of mixtures of transfer agents. The rate at which one transfer agent becomes incorporated in polymer is quite independent of the occurrence of other types of transfer in the system.

For some very reactive transfer agents, such as mercaptans, E_1 is less than E_p, but usually the reverse is true. In most cases, therefore, transfer reactions become more significant at high temperatures. Values for $(E_1 - E_p)$ for styrene with various transfer agents are shown in Table 5.1; the data have been selected from collections by Walling (1957) and Bamford, Barb, Jenkins and Onyon (1958).

TABLE 5.1

EFFECTS OF TEMPERATURE UPON TRANSFER DURING THE POLYMERIZATION OF STYRENE

Transfer agent	$(E_1 - E_p)$(kcal/mole)
benzene	14·8
cyclohexane	13·4
n.butyl bromide	11
styrene monomer	7·7
*iso*propylbenzene	5·5
triphenylmethane	5·1
carbon tetrachloride	5
carbon tetrabromide	1·5
n.butyl mercaptan	−0·7

The viscosity of the reaction medium has less effect upon the rate of transfer reactions than upon those of the growth and termination processes. Both growth and radical-displacement involve a large polymer radical and a small molecule, but as the latter reaction usually requires the higher activation energy it is less liable to become diffusion-controlled. The value of k_1/k_p may therefore increase as polymerization proceeds and the viscosity of the medium increases.

Since radical-displacement is a bimolecular process, it should be accelerated by increases of pressure; the propagation step is similarly accelerated, however, and so the transfer constant, being a ratio of velocity constants, should show little change. This effect is found for transfer to monomer during the polymerization of allyl acetate. The effect of pressure on the efficiency of re-initiation is complicated by the fact that the rates of the wastage reactions may be reduced by the increasing viscosity of the reaction mixture (see Section C of this Chapter).

B. Reactivities in Transfer

Experimental studies of transfer lead only to a value for the ratio k_1/k_p; even if this ratio is known accurately, there may be uncertainty about the precise value of the separate velocity constants. When considering a series of transfer agents with a particular monomer, any doubt concerning the exact value of k_p does not affect the relative values for the velocity constants for the various radical-displacements. For comparisons of the reactivities of various polymer radicals towards a particular transfer agent, however, it is necessary to have absolute values for k_1; in most cases, these are reliable enough to permit firm conclusions to be drawn concerning their dependence upon the natures of the polymer radical and the molecule of the transfer agent. The range of values of k_1 is very wide, as shown by Table 5.2; the values of k_1 are based on values of k_p selected by Bamford and White (1956).

Table 5.2

radical-displacements at 60°C

Polymer radical	k_1 in mole^{-1} l.$^{+1}$ sec^{-1} for reaction with: toluene	n.butyl mercaptan	triethylamine
styrene	$2 \cdot 2 \times 10^{-3}$ (*a*)	3,900 (*b*)	0·12 (*c*)
methyl methacrylate	$7 \cdot 3 \times 10^{-3}$ (*d*)	245 (*b*)	0·30 (*c*)
methyl acrylate	0·565 (*c*)	3,300 (*b*)	83·6 (*c*)
vinyl acetate	7·73 (*e*)	178,000 (*b*)	137 (*c*)

(*a*) Gregg and Mayo (1947), (*b*) Walling (1948), (*c*) Bamford and White (1956), (*d*) Basu, Sen and Palit (1952), (*e*) Palit and Das (1954).

As indicated already, radical-displacement with a transfer agent may occur in more than one way; there are three general methods for identifying the preferred path in the reaction, viz.

(*a*) use of model radicals with the transfer agent and identification and determination of the products;

(*b*) study of the effect of structural modifications in the transfer agent upon its reactivity;

(*c*) in the case of hydrogen-abstractions, examination of the effects of isotopic substitution upon the value of the transfer constant.

Method (*a*) requires no comment. Method (*b*) can be illustrated by considering the transfer constants of a series of aromatic hydrocarbons. In the polymerization of styrene at 60°C the transfer constants for benzene, toluene, *iso*propylbenzene, triphenylmethane and *tert.*butylbenzene are in the ratio 1:7:37:46:194:3·3 (Gregg and Mayo, 1947). Evidently for these hydrocarbons, benzylic hydrogen atoms are the most susceptible to transfer; the reactivity in transfer is raised by increasing substitution at the α-carbon atom. The first stage in transfer to *iso*propylbenzene, for example, is therefore written as

$$P\cdot + C_6H_5.C(CH_3)_2H \longrightarrow P.H + C_6H_5.C(CH_3)_2\cdot$$

although this is not the sole reaction.

The very high reactivity as transfer agents of all compounds containing the SH group is a strong indication that the first stage in transfer can be represented as

$$P\cdot + R.SH \longrightarrow P.H + R.S\cdot \qquad (9)$$

Many disulphides are quite powerful transfer agents. The contrast between the high reactivity of these substances and the low reactivity of monosulphides suggests that the first stage in transfer to a disulphide is not hydrogen-abstraction but rather attack on the S—S bond thus:

$$P\cdot + R.S.S.R \longrightarrow P.S.R + R.S\cdot$$

Some confirmation of this view is provided by studies of cyclic disulphides (Stockmayer, Howard and Clarke, 1953; Tobolsky and Baysal, 1953). When styrene or vinyl acetate is polymerized in the presence of diethylether disulphide, the resulting polymer contains several sulphur atoms per molecule; it is thought that the incorporation of the sulphide in polymer occurs thus:

$$P\cdot + \overline{S.CH_2.CH_2.O.CH_2.CH_2.S} \longrightarrow P.S.CH_2.CH_2.O.CH_2.CH_2.S\cdot$$

If a hydrogen atom in a transfer agent is replaced by deuterium, there is a significant reduction in the observed transfer constant only if the hydrogen atom at that particular site is involved in the abstraction reaction; it arises from the greater strength of the bond involving the heavy isotope. By selective labelling of a molecule, it is possible to locate the hydrogen atoms most likely to react. The transfer constant for $C_4H_9.SD$ is only about a quarter of that for $C_4H_9.SH$ (Wall and Brown, 1954), confirming that hydrogen abstraction is largely according to (9). Other examples of the use of the isotope effect will be given in connection with transfer to monomer during the polymerization of allyl acetate, and with the action of certain retarders (see Chapter 7, C).

Gregg and Mayo (1947) pointed out that for a series of related transfer agents with a particular polymer radical, a reduction in the activation energy for hydrogen-abstraction is frequently accompanied by a reduction in the frequency factor, A (see Fig. 5.1).

A relationship of this type is found for reactions of many types. If the structure of one of the reactants is modified in such a way that the activation energy is lowered, more stringent conditions are imposed on the configuration of the transition state and A is thereby reduced. In studying a series of radical-displacements, it is advisable to determine the values of E and A for the reactions, and not to rely on comparisons of the velocity constants at a single temperature.

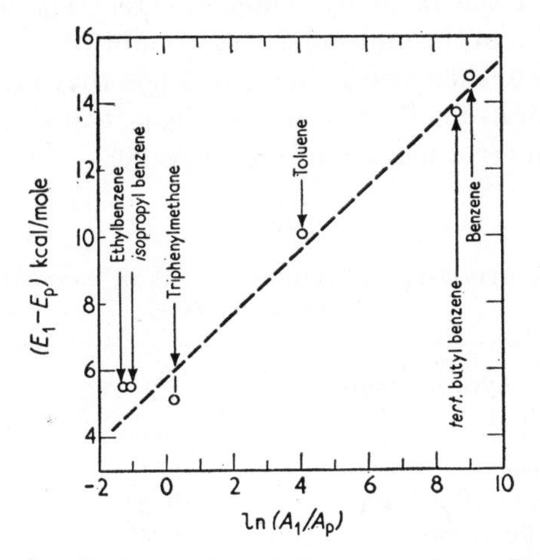

FIG. 5.1. Relationship between A and E for reactions of polystyrene radicals with various hydrocarbons.

In the simplest scheme for correlating data on the rates of hydrogen-abstractions, the velocity constant for the reaction

$$P\cdot + H.B \longrightarrow P.H + B\cdot \tag{10}$$

is supposed to be governed only by the reactivity of the polymer radical and the strength of the bond H—B; these in turn are related to the strength of the bond H—P and the reactivity of the radical B•. The reaction occurs most readily when the radical P• is reactive and the bond H—P strong, and the radical B• is unreactive and the bond H—B weak. On this basis, it would be expected that transfer would always be more frequent in the polymerization of vinyl acetate than in that of styrene, since co-polymerization studies indicate that the polymer radical derived from the former is the more reactive.

This simple approach is adequate for hydrocarbon transfer agents. Bamford, Jenkins and Johnston (1959*b*) compared the values of k_{10} for the reactions of various polymer radicals with a number of hydrocarbons. The velocity constant for hydrogen-abstraction could be expressed as the product of two terms, the one governed by the nature of the polymer radical, and the other by that of the transfer agent. The reactivity of the polymer radical was characterized by the velocity constant, k_T, for the reaction by which the polymer radical abstracts a hydrogen atom from toluene, so that

$$k_{10} = \gamma . k_T \tag{11}$$

where γ is a measure of the reactivity of the hydrocarbon and is related to the strength of the bond by which the reacting hydrogen atom is attached to the rest of the molecule. A reasonably consistent set of values for γ could be established (see Table 5.3); when they are used in conjunction with values of k_T for the polymer radicals, transfer constants can be predicted with an error usually of less than 20%.

TABLE 5.3

REACTIVITIES OF HYDROCARBONS IN TRANSFER REACTIONS AT 60°C

hydrocarbon	γ
benzene	0·15
*tert.*butylbenzene	0·24
cyclohexane	0·42
toluene	1·00 (standard)
*n.*heptane	3·0
ethylbenzene	4·2
diphenylmethane	18·4
triphenylmethane	28·0
pentaphenylethane	16×10^4

This treatment cannot be extended to transfer agents of other types. From data in Table 5.2, γ for *n.*butyl mercaptan would range from $1{\cdot}17 \times 10^6$ (using the polystyrene radical for reference) to $5{\cdot}84 \times 10^3$ (methyl acrylate), and γ for triethylamine from 450 (styrene) to 17·7 (vinyl acetate).

In a hydrogen-abstraction just as in a radical addition process, polar forms may contribute to the structure of the transition state; they would stabilize it and so cause a reduction in the activation energy necessary for the reaction. The effect would be pronounced if the polymer radical is of a type which may readily donate electrons, and the transfer agent one which can readily accept them, or vice versa.

The first attempt (Fuhrman and Mesrobian, 1954) to place the ideas of polar effects in transfer reactions upon a semi-quantitative basis was to apply the Q and e treatment used quite successfully in co-polymerizations. The velocity constant for the radical-displacement is then expressed as

$$k = P_r . Q_t \exp(-e_r . e_t)$$

where P_r and Q_t refer to the reactivities of the polymer radical and the transfer agent, and e_r and e_t are measures of the charges on the radical and the transfer agent respectively. The values of P_r and e_r are the same as those used for the radical in growth reactions; values for Q_t and e_t have been derived for a number of transfer agents (Katagiri, Uno and Okamura, 1955; Katagiri and Okamura, 1955). The treatment is moderately successful, but it is open to the very same objections that can be raised in connection with its application to growth reactions, in particular the implication that the polymer radical and the transfer agent carry permanent charges.

Bamford, Jenkins and Johnston (1959*b*) showed that the velocity constant for a radical displacement could be related to k_T, the velocity constant for the reaction by which the polymer radical abstracts a hydrogen atom from toluene, by the equation

$$\log k = \log k_T + \alpha\sigma + \beta \qquad (12)$$

where α and β are constants for a given transfer agent and σ refers to the polymer radical. For hydrocarbons, α is zero and (12) then reduces to (11). The first term on the right-hand side of (12) is a measure of the reactivity of the polymer radical; the third term is governed by the dissociation energy of the bond to be broken in the molecule of the transfer agent. The second term is a measure of the extent to which polar contributions to the transition state lower its energy; its magnitude depends upon both the radical and the molecule involved in the reaction. The value of σ for a polymer radical $P.CH_2.CXY\cdot$ is the algebraic sum of the Hammett para-substituent constants for X and Y; these constants express the powers of substituents to modify the distribution of electrons at the site of reaction. The scale of α-values is based on zero values for hydrocarbon transfer agents; this is a very reasonable basis since polar effects are likely to be at a minimum for reactions involving these substances. The values of α and β for a transfer agent can be deduced from the transfer constants for two polymer radicals with the agent, provided that k_T and σ for the radicals are already known. Values for the velocity constants for abstraction of hydrogen from toluene by certain polymer radicals were given in Table 5.2; values of σ for polymer radicals and of α and β for transfer agents are shown in Table 5.4. From data such as these, the velocity constant for a radical-displacement can be derived, usually within a factor of three; such predictions must be regarded as very satisfactory in view of the enormous range of values which may be covered.

Transfer agents having negative values for α are those which tend to accept an electron from the polymer radical in the transition state. The treatment summarized above is quite satisfactory for these reagents, but it is less so for transfer agents, such as triethylamine, which act as electron-donors and have positive values for α. It was suggested that for the latter type, it might be better to use revised σ-values for the polymer radicals; the possibility of needing special σ-values when applying the

TABLE 5.4

PARAMETERS FOR CALCULATING TRANSFER CONSTANTS AT 60°C

Polymer radical	σ	Transfer agent	α	β
styrene	−0·01	benzene	0	−0·82
methyl methacrylate	0·28	*iso*propylbenzene	0	0·75
vinyl acetate	0·31	chloroform	−1·4	0·9
methacrylonitrile	0·49	carbon tetrabromide	−4·3	5·25
methyl acrylate	0·45	*n*.butyl mercaptan	−4·8	6·05
acrylonitrile	0·66	ferric chloride	−5·65	7·4

Hammett equation to reactions of certain types has been discussed and justified by Jaffé (1952) in his review of the applications of the equation.

In transfer reactions involving hydrocarbons, polar effects are of little significance and k_{10} is proportional to k_T. The treatment implies that $(E_T - E_{10})$ is constant for all radicals, but data are insufficient for this to be tested. The relationship cannot hold if one of the radicals being considered is very reactive, having such a small value for E_T that for some transfer agents E_{10} would have to be negative. There are strong indications that this limitation applies to some reactions of the methyl radical.

Another indication of the importance of polar factors in influencing the rates of radical-displacements is provided by the observation that the reactivities of certain polymer radicals may be modified by complexing with an ion (Bamford, Jenkins and Johnston, 1958). The presence of lithium chloride during the polymerization of acrylonitrile increases k_p but also the velocity constants for radical-displacements with carbon tetrabromide and triethylamine. It is thought that complexes represented as

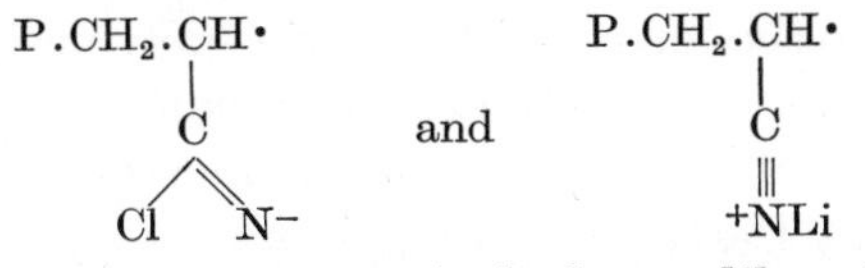

are formed; the former reacts particularly readily with monomer and carbon tetrabromide, and the latter is more reactive than the ordinary polymer radical towards triethylamine.

Polar effects in hydrogen-abstractions are, of course, not confined to reactions involving polymer radicals; relevant data on the reactions of various small radicals have been collected (Trotman-Dickenson, 1959). The existence of polar effects is sometimes deduced from finding that the Hammett equation can be applied to the rates of the reactions of a particular radical with a series of aromatic substrates, one example being the reactions of aralkyl peroxy radicals with para-substituted cumenes (Russell, 1956):

$$R.O.O\cdot + p.X.C_6H_4.CH(CH_3)_2 \longrightarrow R.O.O.H + p.X.C_6H_4.C(CH_3)_2\cdot$$

The nature of the substituent X controls the rate at which the hydrogen atom is abstracted from the *iso*propyl group, and the Hammett equation is satisfied.

C. Transfer to Monomer

If monomer or other unsaturated substance participates in a transfer reaction, the polymer radical may either gain or lose a hydrogen atom:

$$P.CH_2.CHX\cdot + CH_2{:}CHX \longrightarrow P.CH_2.CH_2X + CH_2{:}CX\cdot$$
$$P.CH_2.CHX\cdot + CH_2{:}CHX \longrightarrow P.CH{:}CHX + CH_3.CHX\cdot$$

Depending on the structures of the radical and monomer, other equations can be constructed. The experimentally determined velocity constant, k_f, may therefore be a composite quantity.

In a binary co-polymerization, there are four types of radical-monomer interaction, and for each of them transfer may occur in more than one way; if there are effects due to non-terminal monomer units in the polymer radicals, the number of types of interaction is further increased. It is evident that the system is very complex; even in the simplest cases, it is not possible to determine the velocity constants for the cross-transfers although their sum may be found. Bamford, Barb, Jenkins and Onyon (1958) have concluded that, like cross-propagation and cross-termination, cross-transfer may be favoured. If it occurred readily, it would seriously affect determinations of rates of initiation in co-polymerizations by the molecular-weight method. At this point, it might be mentioned that the action of an added transfer agent in a co-polymerization is, in principle, quite simple. If non-terminal group effects can be ignored, the transfer reactions occurring in a co-polymerization are identical with those taking place in the corresponding homo-polymerizations. The effect of the transfer agent can, therefore, be predicted from the values of k_1/k_p found from examinations of the polymerizations of the separate monomers (Burnett, 1954).

Transfer to certain monomers is accompanied by retardation, the effect being very pronounced during the polymerizations of allyl compounds. The polymerization of allyl acetate is first order with respect to initiator, suggesting that the rate-determining step in the termination process is one involving single radicals; further, the degree of polymerization of the product is low and independent of the rate of polymerization (Bartlett and Altshul, 1945). These results can be explained if transfer to monomer occurs very readily, and if the product radical is so stabilized by resonance that it does not re-initiate polymerization. By studying isotope effects arising from deuteration of the monomer, Bartlett and Tate (1953) showed that the hydrogen-abstraction is largely according to the equation

$$P\cdot + CH_2{:}CH.CH_2.O.CO.CH_3 \longrightarrow P.H + CH_2{:}CH.CH(O.CO.CH_3)\cdot$$

The deuterated monomer $CH_2{:}CH.CD_2.O.CO.CH_3$ polymerizes about twice as quickly as ordinary allyl acetate, giving a product with about double the degree of polymerization.

Increasing the pressure causes the rate of polymerization of allyl acetate to increase, and the order with respect to initiator to change towards 0·5, but there is only a small increase in the degree of polymerization (Walling, 1957). Evidently, k_f/k_p is almost independent of pressure, but the efficiency of re-initiation by the allylic radical rises with increasing pressure. Both re-initiation and the wastage reactions of the allylic radicals are associative processes, and so should be accelerated by application of high pressures. The wastage reactions are, however, inter-radical reactions requiring only small energies of activation; they can become diffusion-controlled, and so, as the viscosity of the medium increases because of the effect of pressure, the balance in the competition shifts to favour re-initiation.

If a hydrogen atom is abstracted from the methyl group in an α-methyl-substituted monomer, $CH_2{:}CX.CH_3$, the product radical can be stabilized by resonance between the structures

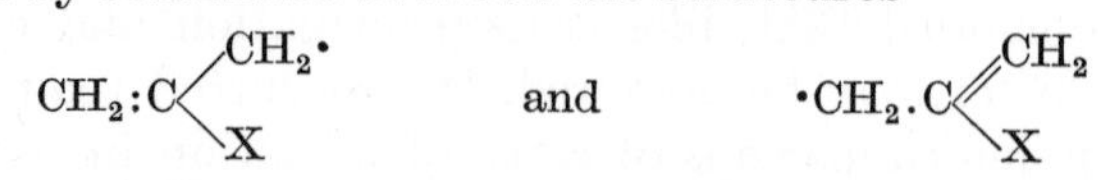

A radical so stabilized would be somewhat unreactive and perhaps inefficient in re-initiation; transfer to monomer during polymerization of

a monomer of this type might therefore be accompanied by retardation. This effect is found in the polymerization of *iso*propenyl acetate $CH_2{:}C(CH_3).O.CO.CH_3$. In other cases, fairly high reactivity of the monomer may make up for low reactivity of the radical derived from it by hydrogen-abstraction, so that the efficiency of re-initiation is quite high; this may be so in the polymerization of methyl methacrylate since there is no evidence for retardation accompanying monomer transfer, although a resonance-stabilized radical might be formed.

D. Transfer to Polymer

As polymerization proceeds and polymer accumulates in the system, there is an increasing possibility of transfer reactions between growing radicals and dead polymer molecules according to an equation such as

$$P\cdot + P.CH_2.CHX.P \longrightarrow P.H + P.CH_2.\underset{\bullet}{C}X.P \qquad (13)$$

The resulting radical may take up monomer molecules to give a branched polymer molecule. From study of the shapes of polymer molecules, it might be possible to make deductions about the frequency of transfer to polymer. Differences between the degrees of branching of samples of polymer can be detected by examination of solution properties of the samples, provided that the branches are not very short; for absolute evaluation of branching, however, it is necessary to have polymers of known structures as standards.

Methods have been devised for synthesizing samples of polyvinyl acetate with known numbers of branches of specified lengths (Melville, Peaker and Vale, 1958*a*, *b*) to act as standards. Most of the methods involve chemical treatment of the polymers and the consequent possibility of considerable changes in the molecular-weight distributions; these changes could of themselves interfere with the interpretation of solution properties. If fractionated materials are used, it must be noted that fractionation may be governed not only by the size of the molecules but also by their shapes, so that the various fractions may have different degrees of branching.

If measurements of branching are used in connection with determination of the rate of transfer to polymer, it must be recognized that branches may grow as a result of reactions of other types also. Unreacted double bonds in a polymer may be involved in growth reactions; these double bonds may be of the type found in polymers of dienes, or they may be terminal groups formed as a result of termination by disproportionation or transfer to monomer.

It is usually assumed that the transfer constant of a monomer unit in a polymer chain is close to that of a model substance, chosen so that the polymer chains attached to the monomer unit are represented by hydrogen atoms or methyl groups. On this basis, ethylbenzene and *iso*propylbenzene are regarded as models for the styrene unit in the polymer, and ethyl acetate and *iso*propyl acetate may represent the vinyl acetate monomer unit. The esters

$$\underset{\displaystyle CO.O.CH_3}{CH_3.\overset{}{CH}.CH_3} \qquad CH_3.\underset{CO.O.CH_3}{CH}.CH_2.\underset{CO.O.CH_3}{CH}.CH_3 \qquad CH_3.\underset{CO.O.CH_3}{CH}.CH_2.\underset{CO.O.CH_3}{CH}.CH_2.\underset{CO.O.CH_3}{CH}.CH_3$$

have been examined as transfer agents in the polymerization of methyl acrylate (Lím and Wichterle, 1958); it was concluded that these molecules can be considered as fair models for one, two or three successive monomer units in the polymer chain.

This treatment is adequate provided that the polymer molecules do not possess a few abnormal groups which are particularly reactive. Such groups might be formed during the preparation of the polymer, or they might be introduced subsequently by chemical modification. Schulz, Henrici and Olivé (1955, 1956) showed that the reactivity of polymethyl methacrylate as a transfer agent may depend upon its molecular weight; the effect can very reasonably be attributed to reactive end-groups. Similarly, polymethyl methacrylate having amine end-groups is much more reactive as a transfer agent with polyacrylonitrile radicals than is a polymer without these end-groups (Bamford and White, 1958). It has also been shown that the reactivity of a polymer in transfer can be increased very considerably by introduction of a few mercaptan groups; the reactivities of such groups are comparable with that of SH in a small molecule (Gluckman, Kampf, O'Brien, Fox and Graham, 1959).

Another approach to the problem of transfer to polymer is to polymerize a monomer in the presence of pre-formed polymer; transfer to polymer followed by re-initiation leads to the production of branched molecules. The weight fraction of the branches in the product can be determined analytically if the branches and the backbone are distinct chemically (Carlin and Shakespeare, 1946); alternatively, isotopic labelling of monomer or polymer can be used (Bevington, Guzman and Melville, 1954; Gleason, Miller and Sheats, 1959). Values for the transfer constants to dead polymer can be derived; the values are not precise because of the very involved procedure and the great experimental difficulties, but it appears again that these constants are probably quite close to those for the appropriate model substances.

Branching by the mechanism just described must be of little importance in polymerizations carried to very low conversions; under these conditions, the concentration of dead polymer in the system is always low and the material is exposed to attack by radicals only for a short time. In this connection, study of the hydrodynamic behaviour of samples of polymethyl methacrylate indicates that branching becomes more important at high conversions (Eriksson, 1956).

Hydrogen-abstraction ordinarily requires an activation energy greater than that for the growth reaction, so that branching should be least pronounced in polymers prepared at low temperatures. The frequency of branching must vary from one system to another, depending upon the reactivity of the growing polymer radical and the strengths of the bonds in the polymer. Branching is expected to be less frequent during the polymerization of styrene than during that of vinyl acetate, because the former monomer gives rise to the less reactive polymer radical.

Simple calculations indicate that ordinarily branching is rare in polystyrene. Suppose that

$$\frac{\text{rate of transfer to polymer}}{\text{rate of consumption of monomer}} = b = \frac{k_{13}[\mathrm{P}\cdot][\text{monomer units in polymer}]}{k_p[\mathrm{P}\cdot][\mathrm{M}]}$$

Consider a system in which 10% of the monomer has polymerized so that

$$[\text{monomer units in polymer}] = [\text{monomer}]/9$$

If k_{13}/k_p is about 9×10^{-5}, as for the appropriate model substances at 60°C, b is 10^{-5}. If the average degree of polymerization is 10^3, at this point in the polymerization polymer molecules are being produced at a hundred times the rate at which branches are being formed. The conclusion that branches are normally rather rare in polystyrene is confirmed by comparison of the solution properties of various samples of polymer. Wall and Brown (1954) could find no difference between the solution properties of polymers prepared under comparable conditions from ordinary styrene and the α-deuterated monomer. Deuterium-abstraction would be expected to be less frequent than hydrogen-abstraction because of an isotope effect; the similarity of the polymers indicates that branching is of negligible importance.

Samples of polyvinyl acetate may differ significantly in their solution properties, and for this material branching may be quite important. Many of the branches in polyvinyl acetate can be detached from the main polymer chain during hydrolysis of the polymer, suggesting that they are attached through an ester linkage, and that the hydrogen-abstraction is largely according to the equation

$$\mathrm{P}\cdot + \mathrm{P.CH_2.}\underset{\displaystyle \mathrm{O.CO.CH_3}}{\overset{|}{\mathrm{CH}}}\mathrm{.P} \longrightarrow \mathrm{P.H} + \mathrm{P.CH_2.}\underset{\displaystyle \mathrm{O.CO.CH_2}\cdot}{\overset{|}{\mathrm{CH}}}\mathrm{.P}$$

The value of $\bar{M}_w/D^2$, where $\bar{M}_w$ is the weight-average molecular weight, and D is the mean extension of the polymer chain in solution, should for a particular polymer be independent of the molecular weight provided that all specimens have the same degree of branching; the higher the frequency of branching, the larger is this quantity. The minimum value of $\bar{M}_w/D^2$ found for any sample or fraction of polyvinyl acetate is about 0·38 (Bosworth, Masson, Melville and Peaker, 1952); this value probably corresponds to polymer without long branches. As the temperature of polymerization is reduced, the value of the ratio tends towards this minimum value, in agreement with the belief that at low temperatures transfer to polymer is very infrequent.

If branch formation is very frequent and pairs of polymer radicals finally combine, insoluble cross-linked polymer should result; such materials are, however, rarely produced during the polymerizations of monomers containing only one carbon–carbon double bond per molecule. Generally, branching is so rare that cross-linking could become evident only at very high conversions; under these conditions, the viscosity of the reaction medium becomes high and the interaction of polymer radicals may be diffusion-controlled so that other reactions, such as transfer to small molecules, become relatively more important. Methyl acrylate is a special case in that insoluble polymers can be produced in the so-called pop-corn polymerization which may occur at temperatures below 50°C. The mechanism of this reaction has been discussed by Bamford, Barb, Jenkins and Onyon (1958); it seems to depend upon immobilization of polymer chains in gels and the consequent trapping of radicals.

During the radical polymerization of ethylene, long branches can develop from intermolecular transfer as just described. This type of branching affects the solution properties of the polymer and certain of its mechanical properties; its frequency increases with increasing conversion in the polymerization and with rising reaction temperature. Much shorter branches, probably containing only four carbon atoms, can also be formed during the polymerization (Roedel, 1953); their number can be determined by measurements of the intensity of the infra-red absorption due to methyl groups. These branches have little effect on the rheological properties of the polymers, but influence properties in the crystalline state. The number of short branches increases with rising temperature in the polymerization, but significant numbers are formed even at very low conversions (Morrell, 1956), so it appears that transfer to dead polymer is not responsible for their growth. The short branches are believed to result from an intramolecular hydrogen transfer, with the transient formation of a cyclic structure:

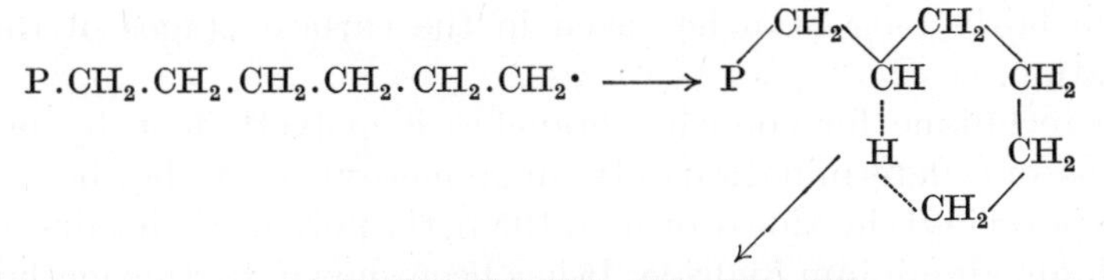

The reaction is commonly known as "back biting".

Short branches are thought to occur also in polyvinyl acetate (Melville and Sewell, 1959). Hydrolysis of a polymer, believed from its solution properties and method of preparation to contain very few long branches, causes a considerable reduction in the average chain length; the molecular weight of the hydrolysed material is almost independent of that of the original polymer, unless it is rather low. It appears that in the main chains of these polymers, there are ester linkages occurring at intervals which depend upon the temperature of polymerization. These linkages could result from an intramolecular transfer process involving the ester group

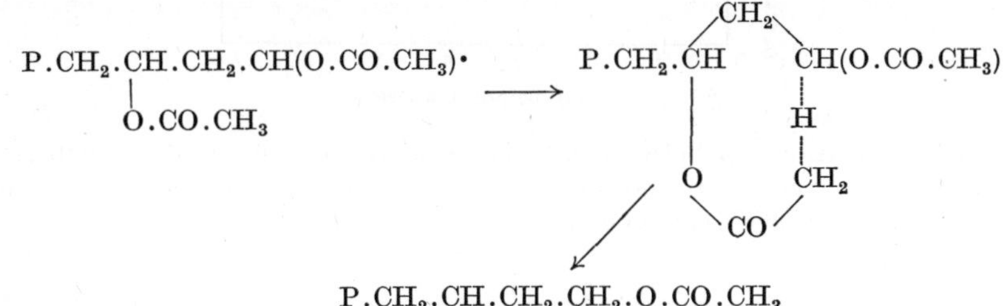

The branches produced by this reaction are so short that they have no appreciable effect upon the solution properties of the polymer.

Melville and Sewell also examined the fractions into which a sample of high-conversion polyvinyl acetate could be split. Study of the viscosities of solutions showed that the various fractions differed in respect of frequency of long branches, the fractions of highest intrinsic viscosity being the most highly branched. Hydrolysis followed by re-acetylation converted the branched fractions into materials having the characteristics of linear polymers; the intrinsic viscosities of all fractions, even those which originally contained linear molecules, were reduced by the treatment (see Fig. 5.2). These results agree with those discussed already for whole polymers.

For stereochemical reasons, "back biting" is most probable when the activated state involves a 5- or 6-membered ring; it might, however, occur also at sites remote from the growing end of the polymer radical, to give fairly long branches even in the earliest stages of the polymerization.

The technique for counting branches in polyethylene by means of spectroscopic determination of terminal methyl groups has been applied also to polyvinyl chloride (Cotman, 1953), the polymer being first reduced by lithium aluminium hydride. It has been shown, by this method, that polyvinyl chloride prepared at 45°C is branched, whereas material made at −40°C is essentially linear (George, Grisenthwaite and Hunter, 1958).

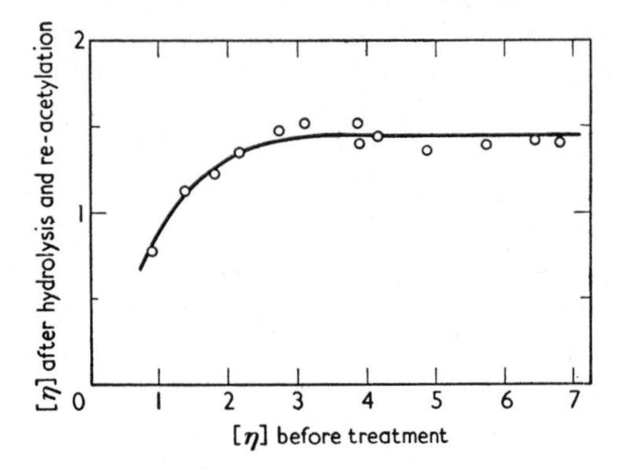

FIG. 5.2. Relationship between intrinsic viscosities of fractions of polyvinyl acetate before treatment and after hydrolysis and re-acetylation—polymer prepared at 18°C to 80% conversion.

E. Transfer to Initiator

Transfer to initiator can properly be regarded as an induced decomposition of that substance since it leads to consumption by a process other than dissociation to radicals. In some cases, the transfer is accompanied by retardation, and this can lead to quite complex relationships between rate of polymerization and [initiator]. Transfer to initiator can be studied by molecular weight methods and also by end-group analysis.

Some of the di-aroyl peroxides are rather susceptible to transfer, and it is for this reason that azo*iso*butyronitrile is commonly preferred as initiator. The radical-displacement most probably can be represented as

$$P\cdot + (R.CO.O)_2 \longrightarrow P.O.CO.R + R.CO.O\cdot \qquad (14)$$

Generally the radical produced in (14), or one derived from it by loss of carbon dioxide, initiates efficiently since it is identical with the radical produced by dissociation of the initiator.

Transfer to initiator leads to abnormal relationships between the rate of polymerization and the kinetic chain length as determined by the initiator fragment method. With brominated dibenzoyl peroxides and styrene, for example, the kinetic chain lengths are low, and the derived rates of initiation correspondingly high compared with values obtained using other initiators (Bevington, Toole and Trossarelli, 1959); the discrepancies become greater as [initiator] is raised. The effect arises from initiator fragments becoming incorporated in the polymer as a result of (14) as well as by the ordinary initiation process. Almost all the radicals formed in (14) become incorporated in polymer, either as a result of re-initiation, or by combination with a growing polymer radical, so that

$$R_{i_{\text{apparent}}} = R_{i_{\text{true}}} + 2k_{14}[\text{P}\cdot][\text{I}] \tag{15}$$

If a labelled di-aroyl peroxide is used, it must be labelled in the aryl group so that incorporation of both types of initiator fragment, aroyloxy and aryl, is measured. The stationary concentration of polymer radicals is given by

$$R_p = k_p[\text{P}\cdot][\text{M}]$$

so that (15) can be changed to

$$R_{i_{\text{apparent}}} = R_{i_{\text{true}}} + \frac{2k_{14}[\text{I}]}{k_p[\text{M}]} \cdot R_p \tag{16}$$

For a fixed concentration of initiator, the rate of polymerization can be varied by adding to the system a second initiator. Then a plot of $R_{i_{\text{apparent}}}$, determined as the rate at which fragments of the first initiator enter polymer, against the rate of polymerization, should be linear; from the slope k_{14}/k_p can be determined, and the true rate of initiation by the first initiator can be found from the intercept (see Chapter 3, C).

This method of working has been applied to transfer reactions between polystyrene, polymethyl methacrylate and polyvinyl acetate radicals and bis(4-methoxy-3,5-dibromobenzoyl) peroxide (Bevington and Lewis, 1960*a*). Applying the treatment of Bamford, Jenkins and Johnston (1959*b*), discussed already in Section B, to the results leads to

$$\alpha = -7 \cdot 57 \qquad \text{and} \qquad \beta = 4 \cdot 96$$

for this peroxide. The large and negative value for α shows that polar factors are of great importance in the radical displacement, and that the form

$$(\text{P}\cdot)^{+} \ (\text{peroxide})^{-}$$

makes a large contribution to the structure of the transition state.

Transfer to a di-aroyl peroxide can affect the relationship between the numbers of aroyloxy and aryl fragments in a polymer. For peroxides susceptible to transfer, the proportion of aroyloxy groups increases with rising concentration of the initiator while [monomer] remains constant (Bevington, Toole and Trossarelli, 1959) (see Fig. 5.3). The aroyloxy radical produced in (14) behaves in the same way as a radical formed by dissociation of the initiator, and may lose carbon dioxide before being captured by monomer. Reaction (14), however, causes aroyloxy radicals to enter polymer without any corresponding increase in the number of incorporated aryl radicals. An increase in the proportion of aroyloxy groups among the combined initiator fragments can also result from primary radical termination (see Chapter 6, E), but that process does not disturb the direct proportionality between R_i, measured by the initiator fragment method, and [initiator].

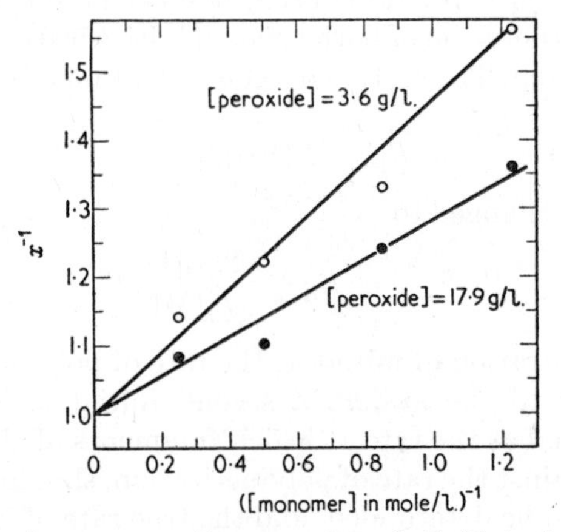

FIG. 5.3. Dependence upon [monomer] and [initiator] of fraction (x) of *m*.bromobenzoyloxy radicals amongst initiator fragments in polystyrene prepared at 80°C.

An older method for studying transfer to initiator requires determinations of the molecular weights of polymer produced in polymerizations with various concentrations of initiator. Most of the available values for transfer constants have been found by this method. According to (7), even when transfer to initiator is significant, the term

$$\frac{k_f}{k_p}+\frac{k_1[\mathrm{AB}]}{k_p[\mathrm{M}]}$$

can be evaluated as the intercept made on the $(\bar{P})^{-1}$ axis by a plot of $(\bar{P})^{-1}$ against $[\text{initiator}]^{1/2}$ for a set of experiments in which [M] and [AB] are constant. If this term is represented as $(\bar{P}_0)^{-1}$, (7) can be rewritten as

$$\frac{(\bar{P})^{-1}-(\bar{P}_0)^{-1}}{[\mathrm{I}]^{1/2}} = \text{constant}+\frac{k_1'[\mathrm{I}]^{1/2}}{k_p[\mathrm{M}]} \tag{17}$$

Plotting the quantity on the left-hand side of this equation against $[\text{initiator}]^{1/2}$, enables k_1'/k_p to be found.

Some of the results of Cooper (1952) concerning substituted dibenzoyl peroxides as transfer agents in the polymerization of styrene, are shown in Table 5.5; these peroxides were studied by the molecular-weight method.

TABLE 5.5

TRANSFER CONSTANTS FOR SUBSTITUTED DIBENZOYL PEROXIDES WITH STYRENE AT 70°C

para substituent	k_1'/k_p	meta substituent	k_1'/k_p
*tert.*butyl	approx. 0	fluorine	0·246
methyl	0·003	chlorine	0·346
methoxy	0·074	bromine	0·465
(hydrogen)	(0·075)	iodine	0·262
fluorine	0·219	cyanide	0·804
chlorine	0·216	nitro	6·2
bromine	0·193		
iodine	0·293		

It is clear from the results for the para-substituted materials that electron-attracting substituents increase the rate of the transfer process. These substituents reduce the rate of the primary dissociation of the peroxide to radicals; to achieve a given rate of initiation with one of these peroxides, it is necessary to use a comparatively high concentration of initiator, and this further favours transfer.

Cooper showed that bis(2,3,4,5-tetrachlorobenzoyl) peroxide engages in transfer quite readily with the polystyrene radical, but less so with polymethyl methacrylate or polyvinyl acetate radicals. This resembles the result already mentioned for bis(4-methoxy-3,5-dibromobenzoyl) peroxide, and again indicates that the peroxide molecule has a strong tendency to be negatively charged in the transition state for the radical-displacement.

Hydroperoxides are very susceptible to induced decompositions and when used as initiators they engage in transfer reactions quite readily. This has been shown very clearly for the *tert.*butyl and cumene compounds with methyl methacrylate (Baysal and Tobolsky, 1952); for a given rate of polymerization, the molecular weights of polymers produced using these peroxides as initiators are very much less than those of polymers prepared using dibenzoyl peroxide or azo*iso*butyronitrile as initiator. Walling and Chang (1954) concluded that transfer to hydroperoxide may involve scission of the O—H bond rather than of the O—O bond.

Laita (1959) has developed another method for studying transfer to initiator suitable for use with photosensitizers. It depends upon the fact that in such a system the rate of initiation can be varied both by altering [sensitizer] and by changing the intensity of the light. Essentially, the method resembles the one involving mixtures of initiators in that the concentration of growing centres can be varied while [initiator] is kept constant. In one set of experiments, the intensity is varied while [sensitizer] is kept constant; the limiting value of $\overline{P}$, as the intensity and the rate of initiation tend to zero, is still governed to some extent by transfer to initiator and from (6):

$$\frac{1}{\overline{P}_{\text{limit.}}} = \frac{k_f}{k_p} + \frac{k_1[\text{AB}]}{k_p[\text{M}]} + \frac{k_1'[\text{I}]}{k_p[\text{M}]} \tag{18}$$

In another set of experiments, the rate of initiation is varied by adjusting [sensitizer]; in this case, the limiting value of $\overline{P}$, as the rate of initiation approaches zero, is unaffected by transfer to initiator and is given by

$$\frac{1}{\overline{P}_{\text{limit.}}} = \frac{k_f}{k_p} + \frac{k_1[\text{AB}]}{k_p[\text{M}]} \tag{19}$$

From the two values of $\overline{P}_{\text{limit.}}$ in (18) and (19), k_1'/k_p can be evaluated.

Laita applied this method to the study of azo*iso*propane; transfer is quite pronounced and may be accompanied by marked retardation. The radical produced in the first stage of the transfer must be less reactive than those formed by dissociation of the initiator. It is possible that the radicals produced in the photolysis are in an excited electronic state and so endowed with special reactivity, but it is more likely that the radical-displacement gives a product quite different from that resulting from the photolysis. It was suggested that the reaction can be represented as

$$\text{P}\cdot + \text{C}_3\text{H}_7.\text{N:N}.\text{C}_3\text{H}_7 \longrightarrow \text{P.H} + \cdot\text{C}_3\text{H}_6.\text{N:N}.\text{C}_3\text{H}_7$$

In accord with this, Laita and Macháček (1959) showed that diphenylpicrylhydrazyl readily abstracts hydrogen from azo*iso*propane by a thermal reaction to give the corresponding hydrazine.

Transfer to initiator is important for many of the sulphur-containing initiators, and is frequently of the degradative type. Many of these compounds are efficient initiators, and evidently the radicals formed in the primary dissociation are different from those produced in radical-displacement. In the case of the tetra-alkyl thiuram disulphides, the radical-displacement may be according to the equation

$$\text{P}\cdot + (\text{R}_2\text{N.CS.S})_2 \longrightarrow \text{P.CS.NR}_2 + \text{R}_2\text{N.CS.S.S}\cdot$$

whereas direct dissociation gives first the radical $R_2N.CS.S\cdot$. Support for this scheme comes from consideration of the effects of tetra-alkyl thiuram monosulphides and the corresponding tetrasulphides upon polymerizations (Ferington and Tobolsky, 1958). Tetramethyl thiuram monosulphide is an initiator but not a retarder, while the tetrasulphide is a retarder but not an initiator. The monosulphide cannot act as a source of the $(CH_3)_2N.CS.S.S\cdot$ radical, which is believed to be unreactive towards monomer; it is unlikely that the tetrasulphide could dissociate in such a way as to give either $(CH_3)_2N.CS.S\cdot$ or $(CH_3)_2N.CS\cdot$ which could act as initiating radicals.

Among other initiators which can engage in degradative transfer are certain disulphides. Ferington and Tobolsky (1958) proposed the equation

$$P\cdot + (C_6H_5.S)_2 \longrightarrow P.C_6H_5 + C_6H_5.S.S\cdot$$

for radical-displacement, the disulphide radical being unreactive; direct dissociation is believed to involve scission of the S—S bond to give reactive radicals. Diazothio-ethers act both as initiators and transfer agents; peculiar relationships between the rate of polymerization and [initiator] (Reynolds and Cotten, 1950) are suggestive of transfer being accompanied by retardation. Diazo-aminobenzene is another initiator which is susceptible to degradative transfer; when it is used with styrene, the rate of polymerization passes through a maximum as [initiator] is increased (Haward and Simpson, 1951).

F. Primary Radical Transfer

Primary radical transfer interferes with the determination of rates of initiation by the initiator fragment method (see Chapter 3, C) because of competition between the reactions

$$R\cdot + M \longrightarrow R.M\cdot \qquad (20)$$

$$R\cdot + A.B \longrightarrow R.A + B\cdot \qquad (21)$$

If the radical $B\cdot$ initiates efficiently, other methods of measuring rates of initiation are unaffected, since they are independent of the chemical nature of the radical which actually initiates the growth of a polymer chain.

In suitable systems, it might be possible to study the competition between (20) and (21) and find k_{21}/k_{20} by comparing the rate at which $R\cdot$ radicals are generated with the rate at which they are incorporated in polymer, or alternatively with the rate at which the product R.A accumulates in the system. Use of a fixed monomer and radical pair would lead to comparative values of k_{21} for the reactions of the radical with various transfer agents. Examples of such competitive studies have been quoted by Trotman-Dickenson (1959). The general procedure is to generate radicals in the absence of monomer but in the presence of two transfer agents; the rates of the two abstraction reactions are compared by means of product analysis. In the study of hydrogen abstraction, the reference transfer agent may contain a labile deuterium atom, and the yields of R.D (formed from the reference transfer agent) and R.H (formed from the transfer agent being examined) are measured.

The method used by Szwarc and his colleagues (see Chapter 3, D.*2*) for comparing the reactivities of monomers towards the methyl radical, depends upon primary radical transfer. The reaction

$$CH_3\cdot + C_8H_{18} \longrightarrow CH_4 + C_8H_{17}\cdot$$

can compete with attack of the radical upon a monomer only if [monomer] is very small; under these conditions only very low polymers are produced. If the conditions are such that high polymer is formed, (21) is of negligible importance compared with (20) unless k_{21}/k_{20} is considerably greater than the transfer constant for the polymer radical.

For some small polymer radicals, the velocity constant for reaction with a transfer agent depends upon the size of the radical (Bengough and Thomson, 1960); this effect is largely due to the influence of the group at the non-reacting end of the radical. The initiating radical in a polymerization has, in general, a structure quite different from that of the growing polymer radical; it is quite possible, therefore, that k_{21}/k_{20} could be very different from the transfer constant for the polymer radical. Consider the polymerization of methyl methacrylate in the presence of triethylamine with azo*iso*butyronitrile as initiator; the transfer constant is $8\cdot3\times10^{-4}$ (Bamford and White, 1956). To estimate the value of k_{21}/k_{20}, suppose that the 2-cyano-2-propyl radical can be represented by the polymethacrylonitrile radical; this is a reversal of the more usual procedure of using small radicals as models for polymer radicals. The velocity constants for the transfer reaction of this polymer radical with triethylamine and its addition to methyl methacrylate are about $0\cdot7$ and 300 mole^{-1} l.$^{+1}$ sec^{-1} respectively (Bamford, Jenkins and Johnston, 1959*b*). The ratio k_{21}/k_{20} for a system containing methyl methacrylate, triethylamine and azo*iso*butyronitrile is therefore estimated as about $2\cdot3\times10^{-3}$, which is about three times the transfer constant for the polymer radicals in this system. This is a case where primary radical transfer might have noticeable effects even in reaction mixtures which could give rise to polymers of fairly high molecular weight.

If the radical produced in (21) is comparatively unreactive, the transfer is of the degradative type, and is accompanied by a reduction in the total rate of initiation as well as in the particular rate of initiation found by the initiator-fragment method. In this case, transfer reactions involving polymer radicals also would inevitably be of the degradative type. Kinetic treatment of such systems is extremely complicated.

Primary radical transfer could affect the analytical method for determining transfer constants. Suppose that the rate of transfer is measured as the rate at which B fragments enter the polymer; for each primary radical transfer step without accompanying retardation, one B fragment would enter polymer. The derived transfer constant would therefore increase as [transfer agent] is increased since primary radical transfer would increase in importance; the true value could be obtained as the extrapolated value at zero concentration of transfer agent (Bevington and Troth, 1960). There is some experimental evidence that primary radical transfer may be significant when carbon tetrabromide is used with styrene and azo*iso*butyronitrile (Bevington, 1960).

There are several examples of primary radical transfer in systems containing rubber and other polyisoprenes as macromolecular transfer agents. The abstraction of α-methylenic hydrogen atoms from these polymers by radicals derived from di-*tert.*butyl peroxide can successfully compete with addition of these radicals to the double bonds in the polymers (Farmer and Moore, 1951). Similar results have been obtained using dibenzoyl peroxide as the source of radicals; abstraction of hydrogen from the polyisoprene occurs to an appreciable extent even if the system contains also monomeric methyl methacrylate (Allen, Ayrey and Moore, 1959).

G. Re-initiation in Transfer Reactions

The phenomenon of degradative transfer has been mentioned already in several connections; it is associated with slowness of the re-initiation step so that some of the radicals produced in the radical-displacement are wasted. Some of the features of degradative transfer involving monomers and initiators have already been discussed. When the efficiency of re-initiation is very low, the substance is better considered as a retarder than as a transfer agent. This is certainly so when the radical formed by the displacement process is so unreactive that it is quite incapable of re-initiation; thus a hydrogen atom can be abstracted easily from diphenylpicryhydrazine to give the unreactive hydrazyl radical (Bevington and Ghanem, 1958)

$$P\cdot + (C_6H_5)_2N.NH.C_6H_2(NO_2)_3 \longrightarrow P.H + (C_6H_5)_2N.\underset{\bullet}{N}.C_6H_2(NO_2)_3$$

The distinction between a retarder and a degradative transfer agent, however, is clearly one of degree only (see Section C of Chapter 7).

Degradative transfer leads to the overall polymerization being of order between 0·5 and 1 with respect to initiator. The efficiency (f) of re-initiation can be related to this order (Allen, Merrett and Scanlan, 1955) by the relationship

$$\text{order with respect to initiator} = \frac{(1-f)}{2}\left(1-\frac{R_p}{R_{p_0}}\right)+0 \cdot 5 \qquad (22)$$

where R_p = rate of polymerization in presence of transfer agent,

R_{p_0} = rate of its absence,

The efficiency is governed by the relative rates of

$$\text{B}\cdot + \text{M} \longrightarrow \text{B.M}\cdot \qquad (23)$$

$$\left.\begin{array}{r} 2\text{B}\cdot \longrightarrow \\ \text{B}\cdot + \text{P}\cdot \longrightarrow \end{array}\right\} \text{inactive products} \qquad \begin{array}{r}(24)\\(25)\end{array}$$

In some cases, (24) can be neglected because the product is so very unstable that it readily reverts to the separate radicals; this is so for the case of triphenylmethane for which the product B_2 would be hexaphenylethane. Kice (1954) has shown that the cross-termination (25) is probably always very much favoured. If (24) is neglected, the principle of competing reactions can be applied to (23) and (25), and

$$f = \frac{k_{23}[\text{B}\cdot][\text{M}]}{k_{23}[\text{B}\cdot][\text{M}]+k_{25}[\text{B}\cdot][\text{P}\cdot]}$$

Since

$$R_p = k_p[\text{P}\cdot][\text{M}]$$

$$\frac{1}{f} = 1+\frac{k_{25}\,.\,R_p}{k_{23}\,k_p[\text{M}]^2} \qquad (26)$$

This equation might be used with (22) to determine k_{23}/k_{25}. Both these velocity constants depend upon the nature of the monomer being used, so that it would not be possible to use the values of the ratio for a series of monomers with a particular transfer agent for comparisons of the reactivities of monomers towards a particular radical. This is very unfortunate, since transfer reactions would otherwise be another source of information concerning the relative reactivities of monomers towards radicals; in many cases, the radicals could be of types which cannot easily be produced by the dissociation of initiators, and which cannot be studied by the methods discussed in Chapter 3, D. Since the B· radicals are generated singly by radical-displacement, there is no problem of geminate recombination when considering the efficiency of re-initiation.

Generally, if a monomer is comparatively unreactive so that k_{23} is small, k_{25} is large because of high reactivity of the corresponding polymer radical; these effects together make for low efficiency of re-initiation. High reactivity of the polymer radical means also that the first stage in transfer occurs readily and appreciable numbers of B· radicals are generated; degradative transfer is therefore most likely to arise in systems in which transfer is pronounced. Table 5.6 shows results quoted by Allen and McSweeney (1958) on transfer reactions involving dihydromyrcene; the transfer agent is unsaturated, but apparently it does not engage in co-polymerization. As expected, the efficiency of re-initiation is low for the monomers which are themselves rather unreactive and which give reactive polymer radicals.

TABLE 5.6

TRANSFER REACTIONS INVOLVING DIHYDROMYRCENE AT 60°C

Monomer	Transfer constant × 10^4	Efficiency of re-initiation
styrene	2	100%
methyl methacrylate	8	100%
acrylonitrile	450	25%
vinyl acetate	730	48%

Similar conclusions can be drawn from studies of triphenylmethane (Bevington and Troth, 1960) and triethylamine (Bamford and White, 1956) as transfer agents. Polyvinyl acetate radicals readily abstract hydrogen from triphenylmethane, but the transfer reaction is accompanied by marked retardation (see Fig. 5.4) and the order with respect to initiator is close to 1·0; on the other hand, retardation is absent when styrene is used as the monomer and the rate of polymerization is very nearly proportional to $[\text{initiator}]^{1/2}$. Triethylamine retards the polymerizations of vinyl acetate, methyl acrylate and acrylonitrile, but not those of the more reactive monomers styrene and methyl methacrylate.

In a similar way, it is possible to relate the reactivity of a transfer agent in a radical-displacement with the reactivity of the derived radical in re-initiation. Thus, very easy abstraction of a hydrogen atom from a hydrocarbon implies that the resulting radical must be rather stable and therefore unreactive. Comparisons of this sort must be restricted to transfer agents of the same type, and it would be wrong to compare a system in which hydrogen is abstracted from a hydrocarbon with one in which an S—H bond is broken; for a given series of transfer agents, however, the ones most reactive in the first stage of transfer are likely to be those least reactive in the second stage. The situation is similar to that found for monomers in co-polymerization, the most reactive monomers giving rise to the least reactive polymer radicals.

Any reduction in the rate of polymerization with increasing concentration of transfer agent is even more pronounced when conditions are such that polymer precipitates during the reaction. Small radicals derived from the transfer agent, unlike the large polymer radicals, can escape out of the precipitate; this means that the acceleration resulting from occlusion of radicals is diminished. This effect can arise even if the transfer agent is not of a type which gives rise to degradative transfer; it is yet

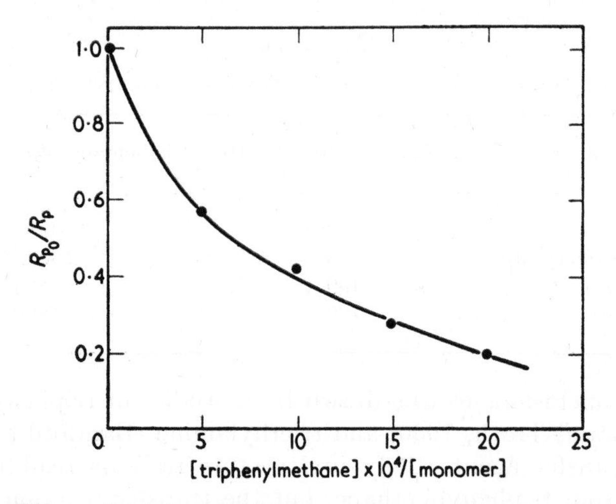

FIG. 5.4. Effect of triphenylmethane upon rate of polymerization of vinyl acetate at 60°C, using azo*iso*butyronitrile at 0·4 g/l. (R_{p_0} = rate in absence of transfer agent).

another reason why caution is needed in interpreting results obtained from systems where radical occlusion can occur.

If solvent is involved in degradative transfer, the kinetic order of the polymerization, with respect to monomer, becomes greater than 1. When [solvent] is high and [monomer] is therefore low, the stationary concentration of growing radicals is depressed because of the occurrence of termination by a process in addition to the normal mutual termination; [P·] is not independent of [monomer], and since

$$\text{rate of consumption of monomer} = k_p[\text{P}\cdot][\text{M}]$$

the order, with respect to monomer, is greater than 1. The solution polymerization of acrylonitrile has been examined in this connection (Thomas, Gleason and Pellon, 1955; Onyon, 1956). Burnett and Loan

(1955) developed a kinetic treatment to explain the dependence of rate of polymerization upon [monomer]. Plots of rate (in fractional units such as %/hour) against [monomer] may have shapes shown in Fig. 5.5. The treatment is attractive since the different shapes can be explained as being due simply to a difference between the efficiencies of re-initiation, this being least for case (*c*). There is, however, an inconsistency in the detailed treatment since the rate of radical-displacement is put equal to the rate of re-initiation. Other kinetic analyses (Kice, 1954; Allen, Merrett and Scanlan, 1955) have been mentioned already. Jenkins (1958)

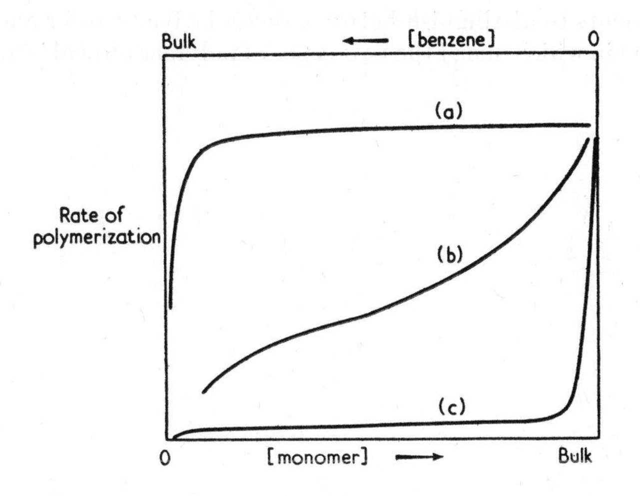

FIG. 5.5. Sketches showing types of dependence of rate of polymerization (in fractional units) upon [monomer] in benzene. (*a*) methyl methacrylate; (*b*) methyl acrylate; (*c*) vinyl acetate.

has shown that transfer constants might be calculated from rate measurements only, if transfer is of the degradative type. His treatment seems best when retardation is pronounced and the efficiency of re-initiation is low, and in such cases reasonable values for transfer constants can be derived; for other systems in which the efficiency is obviously quite high, the derived values for the transfer constants increase continuously as [monomer] is reduced, and evidently the treatment is not satisfactory.

Much of the work on the effects of diluents on the kinetic order of polymerization has been done with benzene or other aromatics. The pronounced retarding effect of benzene upon the polymerization of vinyl acetate might be attributed to the formation of phenyl radicals from the diluent, and their failure to re-initiate; this would conflict with the results of work with dibenzoyl peroxide and vinyl acetate which has shown that these radicals initiate efficiently. It must be recognized that radicals can interact with aromatic substances by mechanisms not yet fully understood. In this connection, the demonstration that benzene can co-polymerize with vinyl acetate (Stockmayer and Peebles, 1953) is of great significance; the retardation is quite likely to be due to inertness of the radical

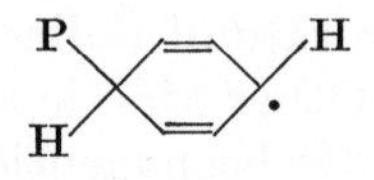

towards the relatively unreactive monomer. It is not possible by rate measurements to distinguish between degradative transfer and co-polymerization in which one of the two types of polymer radical is unreactive.

CHAPTER 6

Interaction of Radicals

A. INTRODUCTION

Reactions by which polymer radicals are deactivated have been mentioned at many points in the preceding chapters, and now special attention will be paid to them. This chapter is concerned with termination reactions involving pairs of radicals; these are commonly referred to as mutual terminations. Processes in which single radicals react with components of the reaction mixture to give relatively unreactive products are considered in Chapter 7; these reactions give rise to the phenomena of inhibition and retardation.

Usually mutual termination causes the steady rate of polymerization to be proportional to (rate of initiation)$^{1/2}$; normally this dependence is revealed by the rate being proportional to [sensitizer]$^{1/2}$ or (intensity of light)$^{1/2}$. If all growing radicals are deactivated singly, the order of the polymerization with respect to initiator becomes unity. It is usually supposed that if this order is between $0 \cdot 5$ and $1 \cdot 0$, mutual and single-radical terminations are both of significance, but it is possible for the order to be greater than $0 \cdot 5$ even if only mutual termination occurs. This effect is found when the growing radicals become embedded in precipitated polymer so that there is a physical barrier to the normal termination process. To some extent, the trapping of radicals in precipitated polymer can be regarded as a process by which they are removed singly from the reaction mixture; the radicals are not deactivated by burial, however, and eventually their reactivity is destroyed by mutual interactions. The subject of diffusion-control in termination processes is considered in Section C, where it is pointed out that it may be of much greater importance than is commonly supposed. In some systems, an appreciable number of polymer radicals may be deactivated by reaction with the small radicals derived from the initiator. This process, known as primary radical termination and discussed in Section E, leads to an order of less than $0 \cdot 5$ with respect to initiator.

In kinetic analyses of polymerizations, there are two distinct conventions concerning mutual termination. The rate of termination is written as $2k_t[\mathrm{P}\cdot]^2$ by some authors, and as $k_t[\mathrm{P}\cdot]^2$ by others. The possibility of this difference must be borne in mind when comparing values of velocity constants derived by different workers.

Section B is concerned with the mechanism of mutual termination, i.e. whether it occurs by combination or disproportionation. The problem has already been mentioned in a number of connections, including the determination of rates of initiation by the molecular-weight method. Even now, there is agreement on the relative importances of the two processes for only a few monomers.

Mutual termination in co-polymerizations is discussed in Section D. The chief point of interest lies in the value of the velocity constant for cross-termination, compared with the values for the velocity constants for the interactions of pairs of similar radicals.

B. Combination and Disproportionation

If the penultimate carbon atoms in polymer radicals carry hydrogen atoms, mutual termination may occur by either combination or disproportionation:

$$2P.CH_2.CXY\cdot \longrightarrow P.CH_2.CXY.CXY.CH_2.P \quad (1)$$

$$\longrightarrow P.CH_2.CHXY + P.CH{:}CXY \quad (2)$$

Another type of disproportionation may be possible in some cases; thus for an α-methyl substituted monomer, such as methyl methacrylate, the hydrogen atom may be transferred from the methyl group:

$$2P.CH_2.C(CH_3)X\cdot \longrightarrow P.CH_2.C(CH_3)XH + P.CH_2.CX{:}CH_2 \quad (3)$$

The relative importances of the various types of mutual termination are likely to depend upon the natures of the radicals and the reaction temperature, and perhaps also upon the sizes of the radicals. In co-polymerizations, further complications may arise. During the co-polymerization of n monomers, there may be $n(n+1)/2$ types of radical interaction, each of which may be by combination or disproportionation; each disproportionation may occur by several different mechanisms, since, in general, either of the radicals may give up a hydrogen atom to the other.

Numerous attempts, summarized by Burnett (1954) for example, have been made to deduce the mechanism of interaction of polymer radicals from study of the behaviour of small radicals, for which the rates of combination and disproportionation can be compared from analyses for the various products. For small radicals, the balance between the alternative reactions depends quite critically upon the structures of the radicals; for extension of the results to polymer radicals, it is essential that the structures of the models should be chosen carefully. Too much reliance must not be placed on conclusions drawn from the use of model radicals in this connection, or in the study of other elementary processes in radical polymerizations. The structures of radicals which can be generated in a controlled fashion by the direct dissociation of molecules seldom correspond to more than one monomer unit, and there may be significant effects of end-groups in modifying reactivity. Studies of model radicals do little beyond confirming that both combination and disproportionation may be possible in polymerizations. It is significant, however, that the energy of activation for the combination of small radicals is close to zero, and that for disproportionation is usually quite small.

The polymethyl methacrylate radical is one of the very few polymer radicals for which the occurrence of appreciable disproportionation has been reasonably well established. This may be due in part to the fact that five hydrogen atoms are available for the disproportionation of polymethyl methacrylate radicals by either (2) or (3), whereas in the polystyrene radical, for example, only two hydrogen atoms can take part in the reaction. Another important factor may be that if two polymethyl methacrylate radicals combine by (1), the product has quite bulky substituents on neighbouring carbon atoms.

An important method for studying the alternative termination processes in sensitized polymerizations involves comparisons of the values of the rate of initiation and the kinetic chain length derived by different methods (see Chapter 3, C). Provided that certain conditions are satisfied, some of the methods give values for the kinetic chain length without any assumptions concerning the mechanism of termination or the frequency of transfer; thus, the initiator-fragment method depends upon initiation being solely by the addition to monomer of a particular radical, and in the method involving knowledge of the rate of decomposition of the initiator, it is necessary only to know the efficiency of initiation. Calculation of the kinetic chain length during a polymerization from measurements of the average molecular weight of the resulting polymer, however, requires a knowledge of the mode of termination and of the importance of transfer reactions. This procedure has been thoroughly discussed by Baysal and Tobolsky (1952) and used recently by Bamford, Jenkins and Johnston (1959*a*) to show that combination predominates during the polymerization of acrylonitrile.

Another approach to the problem of the mechanism of termination requires analysis of the polymer for initiator fragments. An empirical formula

$$(\text{monomer unit})_n(\text{initiator fragment})_1$$

can be derived; it can be converted to a molecular formula

$$(\text{monomer unit})_{nx}(\text{initiator fragment})_x$$

if the number average molecular weight of the polymer is known. The procedure is essentially a comparison of the molecular weight determined by end-group analysis with that found by a method such as osmometry. Both methods give number average values; if the same sample of polymer is used, losses of small polymer molecules during recovery and purification should not affect the value derived for x. If every initiation step introduces a fragment of initiator into polymer and if transfer

TABLE 6.1

COMBINATION AND DISPROPORTIONATION IN THE POLYMERIZATION OF METHYL METHACRYLATE

	$k_{comb.}/k_{disp.}$		
Temp.(°C)	ref. (*a*)	ref. (*b*)	ref. (*c*)
0	0·67	—	—
25	0·44	—	—
40	—	—	2·22
60	0·17	0·74	1·33
80	—	—	0·77

(*a*) Bevington, Melville and Taylor (1954*a*, *b*).
(*b*) Ayrey and Moore (1959).
(*c*) Schulz, Henrici-Olivé and Olivé (1959).

reactions are very infrequent, the limiting cases of exclusive disproportionation and exclusive combination correspond respectively to values of 1 and 2 for x; in general,

$$\frac{k_{combination}}{k_{disproportionation}} = \frac{2(x-1)}{2-x}$$

In the case of styrene with azo*iso*butyronitrile as initiator, disproportionation has been shown to be of very little significance for polymerizations at both 25° and 60°C (Bevington, Melville and Taylor, 1954*a*, *b*); this result is now generally accepted. Disproportionation occurs during the polymerization of methyl methacrylate, however, and its relative importance increases with rising temperature; $(E_{disp.} - E_{comb.})$ is about 5 kcal/mole. Differences between the values of $k_{comb.}/k_{disp.}$ quoted by various workers (see Table 6.1) are very largely due to differences between their interpretations of osmotic measurements for number average molecular weights.

Similar studies of the co-polymerization of styrene and methyl methacrylate at 60°C showed that the cross-termination occurs by combination (Bevington, Melville and Taylor, 1954*b*).

It is necessary to consider the effects of various transfer reactions when dealing with the number of initiator fragments per polymer molecule. Ordinarily, transfer to polymer is very infrequent, and in any case it does not, of itself, affect the average number of initiator fragments per polymer molecule. Transfer to initiator makes the number of fragments tend towards 2 even if termination is by disproportionation. Other types of transfer reduce the average number of initiator fragments per polymer molecule. If allowance is made for transfer, it is found that combination is much the more important mode of termination during the polymerization of acrylonitrile at 60°C (Bevington and Eaves, 1959; Bailey and Jenkins, 1960).

The distribution of molecular weights in addition polymers depends upon the frequency of transfer reactions and the mechanism of the termination process; these factors also govern the relationships between the various average molecular weights. Physical examination of polymers might, therefore, lead to assessment of the relative importances of combination and disproportionation during polymerizations; in practice, the experimental difficulties are great. Only recently have reliable methods of fractionation been developed; by one of them, it has been shown (Baker and Williams, 1956) that the distribution in polystyrene prepared at 60°C conforms almost exactly with that predicted on the assumption of complete combination.

Bamford and Jenkins (1955) have used another approach to the problem of the mechanism of termination. Polymers, prepared using initiators which give radicals containing functional groups, are treated with suitable difunctional reagents which can cause the polymer molecules to link together through the initiator fragments. Azo-nitriles of formula $CH_3.C(CN)(CH_2.CH_2X).N:N.C(CN)(CH_2.CH_2X).CH_3$, where X is COOH or OH, can be used for this purpose. If all the polymer molecules contain two fragments, the treatment should cause the average molecular weight to rise many times. The possibility that polymer molecules may be linked together by reactions involving groups other than those derived from the initiator, can be tested by submitting to the treatment polymers prepared using initiators giving radicals without functional groups. If these tests prove satisfactory, and provided that transfer to initiator during the polymerization is rare, a large increase in the average molecular weight in the main experiment is sufficient to establish that combination must have occurred during the polymerization. Failure to increase the molecular weight substantially is not proof that combination is absent; the average number of initiator fragments per polymer molecule is certainly less than 2 if disproportionation is significant, but it can also result from the occurrence of transfer reactions.

Generally, results obtained by this technique have confirmed those obtained otherwise.

The coupling of polymers containing reactive end-groups has been examined in detail (Bamford and Jenkins, 1960; Bamford, Jenkins and Wayne, 1960), and the relative concentrations of polymer and linking agent needed to produce a large increase in molecular weight have been defined. In addition to its importance in connection with the mechanism of termination reactions, the technique is valuable for the preparation of block co-polymers from mixtures of polymers.

Various other types of experiment give limited information on the problem of the mechanism of mutual termination. One of these involves a comparison of the effect of a retarder upon the rate of polymerization

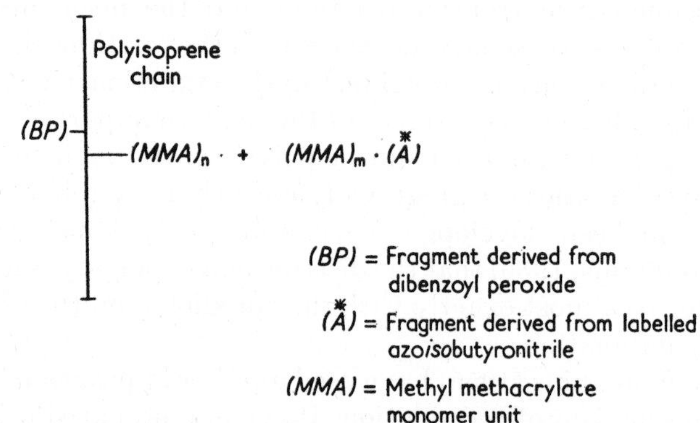

FIG. 6.1. Representation of a termination process in the polymerization of methyl methacrylate in the presence of gutta-percha.

with its effect upon the molecular weight of the product (see Chapter 7, D); a study of this type involving *p*.benzoquinone has shown that disproportionation probably occurs during the polymerization of methyl methacrylate. Another completely different procedure has shown that some combination may occur during this polymerization at 60°C. If methyl methacrylate is polymerized in the presence of gutta-percha with ^{14}C-azo*iso*butyronitrile as initiator, no grafting occurs and the gutta-percha does not acquire any ^{14}C-activity; addition of unlabelled dibenzoyl peroxide to the system, however, causes grafting and the inter-polymer contains fragments derived from the azo initiator (Allen, Ayrey and Moore, 1959). Certain aspects of this work have been discussed already (see Chapter 4, C); its present significance is that the radical interaction represented in Fig. 6.1 must occur, in part at least, by combination.

The fact that polymer radicals of some types are more likely to undergo combination and that others tend to disproportionate, is probably of considerable significance in the radiation chemistry of high polymers. Some polymers, when exposed to high-energy radiations, become cross-linked, while for others the main process is degradation. Cross-linking is likely to occur when the radicals produced during irradiation are of types prone to combination, as in the case of polystyrene. If main-chain scission during irradiation, perhaps followed by rearrangement (see Chapter 4, A.*4*), gives radicals which disproportionate, the break in the polymer chain may be made permanent with the result that degradation is observed. Quite obviously, these are over-simplifications for rather complex systems, but undoubtedly there is a connection between the mechanism of the interaction of polymer radicals and the gross effects of radiations upon a high polymer.

C. Diffusion-control in Termination

Evidence from gas-phase reactions indicates that inter-radical reactions have high collision efficiencies; it is quite likely, therefore, that the rate of mutual termination in radical polymerizations in solution may be governed by the rate at which the large radicals can diffuse through the reaction mixture. Diffusion-control of termination gives rise to the "gel effect" in which, at a certain point in the reaction, both the rate of polymerization and the molecular weight of the product rise markedly because of a reduction in the rate of termination (Norrish and Smith, 1942). It causes a change in the kinetic characteristics of the polymerization which can account for doubly peaked or rather broad distributions (Eriksson, 1956) and for large values of $\bar{M}_w/\bar{M}_n$ for the unfractionated polymer. For some monomers, the gel effect may set in at a very early stage; the rate of reaction may rise continuously from the beginning of the polymerization until the growth reaction also becomes diffusion-controlled. A treatment of diffusion-control of the various elementary steps has been discussed in Section A.*3* of Chapter 4.

The onset of diffusion-control can be delayed by reducing the viscosity of the reaction mixture, either by performing the polymerization in a diluent which is a good solvent for the polymer, or by adding a powerful transfer agent to the system so that the molecular weight of the polymer is reduced appreciably and the solution is less viscous. The gel effect can be enhanced by adding pre-formed polymer to the system so that the viscosity is high even before polymerization has started, or by adding a small proportion of a cross-linking agent to the reaction mixture.

Results of Bengough and Melville (1955) shown in Table 6.2 indicate that k_t for vinyl acetate apparently decreases from the beginning of the polymerization and that E_t rises correspondingly. This suggests that for this monomer, and perhaps for others, true values of the velocity constant for the interaction of radicals have not yet been obtained because at all stages in the polymerization the termination is diffusion-controlled. Examination of the polymerization of methyl methacrylate in a series of solvents of low chemical reactivity and good solvent power (Benson and North, 1959) has shown that, at 40°C, k_t is inversely proportional to the viscosity of the reaction medium over a wide range; for this monomer, therefore, termination may be diffusion-controlled even at the earliest

TABLE 6.2

MUTUAL TERMINATION IN THE POLYMERIZATION OF VINYL ACETATE AT 25°C

% polymerization	$10^{-5} k_t$ (mole^{-1} l.$^{+1}$ sec^{-1})	E_t (kcal/mole)
4	240	<1
23	126	1·4
46	90	3·4
57	6·7	8·6
65	1·15	>13

stages of polymerization in fluid solvents such as ethyl acetate. All the values of k_t quoted for methyl methacrylate may therefore be considerably less than the real velocity constant for the interaction of pairs of polymer radicals. The reactivity of a polymethyl methacrylate radical in termination may also appear to depend upon its size because this affects its diffusivity. If termination is diffusion-controlled at all stages in the polymerization of methyl methacrylate, there must be doubt concerning the explanation of the so-called gel effect in the polymerization of this monomer. Benson (1960) has pointed out that diffusion-control of termination may be very common in radical polymerizations, but it cannot operate in all systems; for example, with *n*.butyl acrylate at 30°C, k_t is almost independent of viscosity when that is low (Benson and North, 1959).

In an ideal polymerization, the order with respect to monomer should be unity (see Chapter 4, A.*1*); one of the explanations for orders a little greater than this, is that the rate of termination is comparatively high at low concentrations of monomer. Under these conditions, the average length of polymer radicals at termination is comparatively small, and such polymer radicals diffuse rather more easily than larger ones. If there is diffusion-control of termination, k_t for the smaller radicals must be larger than that for the larger radicals.

Application of high pressures should increase the rate of chemical interaction of radicals since the transition state is more compact than the reactants, and yet k_t for styrene at 30°C decreases as the pressure is raised (Nicholson and Norrish, 1956)—(see Fig. 6.2). The effect can be explained on the basis of diffusion-control since the viscosity of the reaction mixture increases as the pressure is raised. Even at 3000 kg/cm^2, however, the viscosity of the medium is probably only about 3 centipoises, whereas Benson and North varied the viscosity between about 0·4 and 1000 centipoises by changing the nature of the solvent. It appears, therefore, that the polymerization of styrene may be another reaction much more sensitive to diffusion-control of termination than is commonly supposed.

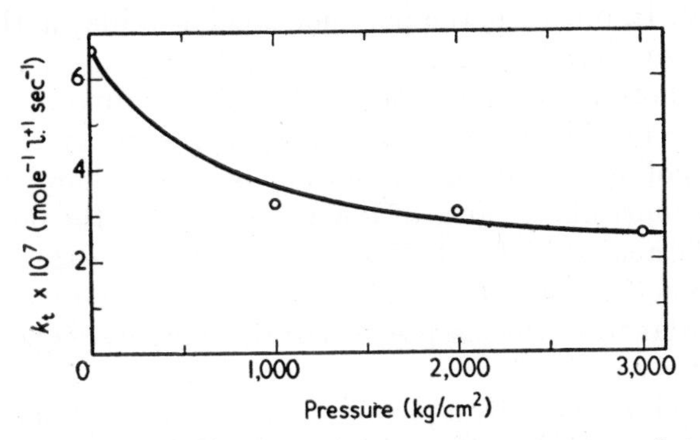

FIG. 6.2. Effect of pressure upon k_t in the polymerization of styrene at 30°C (Nicholson and Norrish, 1956).

Diffusion-control of inter-radical reactions at high pressures is important in another connection also. As explained in Chapter 5, C, it can account for the effect of pressure upon the efficiency of re-initiation during transfer to monomer in the polymerization of allyl acetate.

Diffusion of polymer radicals is hindered in systems where polymer is precipitated as it is formed. The precipitated polymer radicals are most probably tightly coiled and tend to coalesce with other particles of polymer. The reactive ends of the radicals are buried and inaccessible to further reaction. Except for transfer to polymer, all the elementary reactions must be reduced in rate, but the effect is greatest on the bimolecular termination process since it requires the close approach of the reactive ends of two occluded radicals. There are no abnormalities in the actual chemical reactions involved in these polymerizations, and normal kinetic behaviour is found for reactions performed in diluents which are good solvents for both monomer and polymer. The polymerization of acrylonitrile has been much studied in this connection; the results and the evidence for trapping of radicals have been summarized by Bamford, Barb, Jenkins and Onyon (1958).

There may be abnormal features about polymerization in a liquid which, though not a precipitant for the polymer, is a poor solvent. Polymer molecules and radicals in such a solvent are likely to be tightly coiled so that the reactive points are shielded to some extent. Grassie and Vance (1956) suggested that this may be responsible for high values of the activation energies for the elementary reactions in the polymerization of bulk methacrylonitrile; the monomer is a poor solvent for the polymer which may separate as a highly swollen gel during the polymerization. Diffusion-control of termination may have been the cause of low values for k_t in polymerizations initiated by persulphate in non-aqueous media (see Chapter 2, B). In one case, the polymer was insoluble; in the other, the liquid could not have been a good solvent. An explanation of this type seems much more likely than one in which it is supposed that there is electrostatic repulsion between the charged groups at the unreactive ends of the polymer radicals. Repulsion between charged radicals is, however, the most likely cause of low values of k_t in the polymerization of methacrylic acid at high pH (Blauer, 1960). In this case, the monomer units at the reactive ends of the radicals are charged so that the situation is quite different from that in polymerizations involving persulphate.

Emulsion polymerizations are examples of reactions in which the rate of termination is reduced by a physical effect. These polymerizations are quite rapid and yet give products of very high molecular weight; termination occurs less readily in these systems than in homogeneous polymerizations. According to the accepted picture for emulsion polymerization (Harkins, 1947; Smith and Ewart, 1948), growth occurs in a large number of separate small particles. Radicals are generated in the aqueous phase and enter the reaction sites one at a time. The radicals then grow without interruption, although transfer reactions may occur, until a second primary radical enters; then, because of the small size of the particle, termination occurs within a short space of time. The average interval between the entries into a particle of the first and second primary radicals, depends on the rate at which radicals are generated and on the "concentration" of the particles; the interval may be quite long, however, and the molecular weight of the polymer may be limited only by transfer reactions.

D. Termination in Co-polymerizations

During co-polymerizations, there may be termination reactions involving unlike polymer radicals as well as those between pairs of similar radicals. The velocity constant, $k_{t_{ab}}$, for cross-termination may include contributions corresponding to combination and various types of disproportionation. It is not a quantity which can be determined directly, but it may reasonably be related to the velocity constants, k_{t_a} and k_{t_b}, for the interactions of pairs of like radicals. The velocity constant for an inter-radical reaction may be regarded as the product of two terms, each corresponding to the reactivity of one of the radicals involved, so that

$$k_{t_a} = k_{t_a}'^2, \qquad k_{t_b} = k_{t_b}'^2, \qquad k_{t_{ab}} = k_{t_a}'.k_{t_b}'$$

and

$$k_{t_{ab}} = (k_{t_a}.k_{t_b})^{1/2}$$

This relationship is commonly used in kinetic analyses of radical chain reactions in which cross-terminations are possible; it reduces the number of independent velocity constants. An extension is to suppose that

$$k_{t_{ab}} = \phi(k_{t_a}.k_{t_b})^{1/2} \tag{4}$$

The rate of a co-polymerization in which all reaction chains are terminated in pairs can be expressed (see Chapter 4, B.*1*) as

$$R_p = \frac{R_i^{1/2}(r_a[\mathrm{M}_a]^2 + 2[\mathrm{M}_a][\mathrm{M}_b] + r_b[\mathrm{M}_b]^2)}{(r_a^2\,\delta_a^2[\mathrm{M}_a]^2 + 2\phi r_a r_b\,\delta_a\,\delta_b\,[\mathrm{M}_a]\,[\mathrm{M}_b] + r_b^2\,\delta_b^2[\mathrm{M}_b]^2)^{1/2}} \tag{5}$$

The "normal" value of ϕ may be 1 or 2, depending upon the way in which k_t is defined (see Section A). In deriving (5), it is assumed that the reactivity of a polymer radical is independent of its size and is governed by the nature of the monomer last added; the equation refers to the instantaneous rate, and not the average rate over a period during which the rate of initiation and the concentrations of the monomers may change significantly. From (5), ϕ can be evaluated provided that the other quantities are known. An alternative method has been proposed (Palit, 1955) but not used to any appreciable extent; it requires only measurements of rates of co-polymerization. It has the limitation that it can be applied properly only to those systems for which there is a well-defined minimum in the plot of rate of co-polymerization against composition of feed, and then the result refers to ϕ at or near the composition corresponding to this minimum.

For pairs of very similar monomers, such as styrene and *p*.methoxystyrene (Bonsall, Valentine and Melville, 1951), ϕ is close to the "normal" value; in other systems, however, it may be considerably greater, and the very high value of 400 has been quoted for the co-polymerization of methyl methacrylate with vinyl acetate (Burnett and Gersmann, 1958). These results indicate that the interaction of unlike radicals in solution may be much favoured over reactions between pairs of identical radicals. Blackley, Melville and Valentine (1954) extended the kinetic treatment to the rate of co-polymerization for a three-component system. Using data obtained from studies of the polymerizations of the separate monomers and of the co-polymerizations of the three binary mixtures, rates of co-polymerization were calculated and compared with the observed values. Results displayed in Table 6.3 show fair agreement between observed and calculated rates, considering just how many experimentally determined quantities are involved in the calculations.

TABLE 6.3

RATES OF CO-POLYMERIZATION IN A THREE-COMPONENT SYSTEM

Concns. of monomers (mole/l.)			Rates of co-polym. × 10^5 ($mole^{+1}$ $l.^{-1}$ sec^{-1})	
methyl methacrylate	styrene	p.methoxy-styrene	observed	calculated
7·34	0·99	0·60	6·45	8·16
5·79	1·66	1·20	5·27	7·26
5·17	1·45	1·82	4·94	5·03
1·44	2·78	1·63	3·66	5·29
2·41	1·65	3·85	4·10	5·38
2·81	2·76	2·60	4·01	7·70

Large values of ϕ are found for co-polymerizations in which there is a pronounced tendency for alternation of monomer units (Melville and Valentine, 1950). Alternation can be attributed to a lowering of the energy of the transition state for the cross-propagation because of contributions of polar forms to the structure of the transition state (see Chapter 4, B.*4*). A similar effect may assist in making ϕ rather large, but it cannot be the sole cause. To account for a value of 100 for ϕ at 60°C,

$$(E_{t_a} + E_{t_b} - 2E_{t_{ab}})$$

must be 6 kcal/mole; activation energies for termination processes are not known accurately, but they are certainly very small and a difference of this magnitude is quite unreasonable. It would be helpful to have information on the dependence of ϕ upon temperature for confirmation.

In some systems, ϕ apparently varies considerably with the composition of the feed; see, for example, Fig. 6.3, which refers to the co-polymerization of *n*.butyl acrylate and styrene (Bradbury and Melville, 1954).

From its definition, which involves only velocity constants, ϕ should be quite independent of the concentrations of the various reactants. The apparent dependence of $k_{t_{ab}}$ upon the composition of the feed is reminiscent of the fact that, for a few co-polymerizations, the monomer reactivity ratios vary similarly. Barb (1953*b*) supposed that the penultimate monomer unit can affect the reactivity of a polymer radical in termination processes, so that the velocity constants for the interactions

$$\left.\begin{array}{l} \mathrm{P.M}_a.\mathrm{M}_a\cdot + \cdot\mathrm{M}_b.\mathrm{M}_b.\mathrm{P} \longrightarrow \\ \mathrm{P.M}_a.\mathrm{M}_a\cdot + \cdot\mathrm{M}_b.\mathrm{M}_a.\mathrm{P} \longrightarrow \\ \mathrm{P.M}_b.\mathrm{M}_a\cdot + \cdot\mathrm{M}_b.\mathrm{M}_b.\mathrm{P} \longrightarrow \\ \mathrm{P.M}_b.\mathrm{M}_a\cdot + \cdot\mathrm{M}_b.\mathrm{M}_a.\mathrm{P} \longrightarrow \end{array}\right\} \text{unreactive products} \quad \begin{array}{r} (6) \\ (7) \\ (8) \\ (9) \end{array}$$

have different values. The relative importances of these processes would depend upon the composition of the feed; at very high values of $[\mathrm{M}_a]/[\mathrm{M}_b]$,

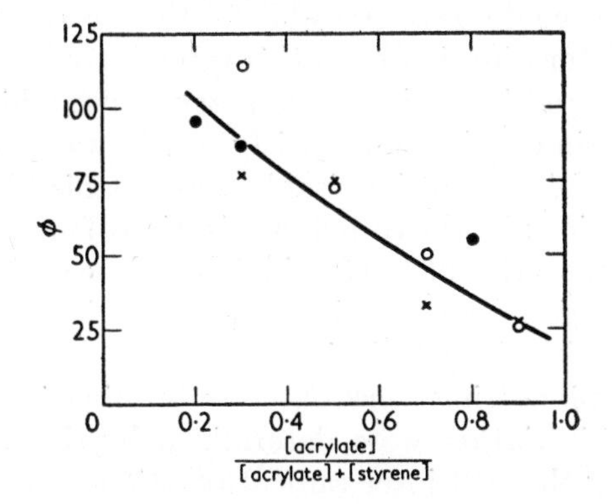

FIG. 6.3. Variation of ϕ with monomer composition for co-polymerization of *n*.butyl acrylate and styrene at 60°C in bulk and in benzene; mole fraction of benzene: ○ 0; × 0·6; ● 0·9.

(7) would be the predominant cross-termination, whereas (8) would be the most important of these reactions when this ratio of concentrations is very low. The treatment has been applied to the cases of *n*.butyl acrylate with styrene (Arlman, 1955) and methyl methacrylate with styrene (Suzuki, Miyama and Fujimoto, 1959). For simplicity, it is assumed that the reactivity of only one of the polymer radicals is affected by the penultimate monomer unit so that

$$k_6 = k_8 \quad \text{and} \quad k_7 = k_9, \quad \text{but} \quad k_6 \neq k_7$$

The treatment can account for the apparent dependence of ϕ upon feed composition, but only by supposing that the penultimate unit reduces the reactivity of the polybutyl acrylate radical by a factor of 10, and that of the polymethyl methacrylate radical by a factor of 3.

Effects of penultimate monomer units upon cross-propagations are rather difficult to detect, and there is no evidence for their existence in the cases under consideration; this, however, does not prove that these effects are absent in the corresponding termination reactions. Propagation is a head-to-tail process, whereas termination is a head-to-head reaction, so that interaction between the penultimate monomer unit in one radical and the last unit in the other radical is perhaps a little more likely. It is doubtful, however, whether it could produce an effect of the size required, on Barb's treatment, to explain the results in the system involving butyl acrylate.

Burnett and Gersmann (1958) pointed out that in those co-polymerizations reported as having high and variable values of ϕ, styrene is one of the monomers. They suggested that a type of degradative transfer to monomeric styrene could be responsible for these results. As [styrene] in the feed is raised, termination from this type of transfer would become more pronounced, and it would be necessary to use a higher value for ϕ to make the experimental results satisfy (5). It was claimed, however, that high values of ϕ are not inevitably accompanied by a dependence of ϕ upon composition; for the co-polymerization of methyl methacrylate with vinyl acetate, a value of 400 for ϕ might fit the experimental results over the whole range of compositions, but there is considerable uncertainty associated with the calculations.

In some of the pioneering work on cross-termination in co-polymerizations, rates of initiation were determined by the molecular-weight method, assuming that all inter-radical reactions occur by disproportionation and that transfer reactions are of negligible importance. Critical re-examination (Melville and Valentine, 1950) of the assumptions showed that it is very unlikely that they could be responsible for gross errors. High values for ϕ could be accounted for if the values used for rates of initiation in the co-polymerizations are too large; variation of ϕ with composition could be explained if the magnitude of the error depends upon the composition of the system. When the initiator-fragment method is used for determining rates of initiation, high values can result from transfer to initiator and primary radical termination. Experiments with single monomers and the common initiators have shown that these processes are not normally of great significance, and it is most improbable that the situation could be very different in a co-polymerization.

The solvent can affect both the energy and entropy of activation for the dissociation of a molecule into radicals because of solvation effects (see Chapter 2, A). The reverse reaction, i.e. the combination of radicals, might be similarly affected; the apparent dependence of $k_{t_{ab}}$ upon the composition of the mixture is, however, too great to be accounted for in this way. Another physical effect of the solvent which may be important is in connection with diffusion. Diffusion-control of termination may be much commoner than has been supposed, and there must, therefore, be doubts concerning the validity of a fairly elaborate kinetic treatment in which velocity constants for the interactions of polymer radicals of various sizes and types are supposed to be independent of the composition of the medium. Inter-radical reactions in the gas-phase are not subject to diffusion-control, and it is significant that high values for ϕ are not found in these systems (Trotman-Dickenson, 1958).

A very serious difficulty in evaluating ϕ concerns the choice of values for δ, i.e. $k_t^{1/2}/k_p$, for the separate monomers. Ideally this quantity should be independent of the composition of the reaction mixture, but, for many monomers, its value depends upon [monomer] during polymerization; this is particularly so for vinyl acetate and *n*.butyl acrylate but it is significant for styrene. The effect is associated with the fact that for the overall polymerization the kinetic order with respect to monomer is greater than 1. It has been indicated at several points in preceding chapters, that one of the most likely explanations is that in the presence of solvent there are additional termination processes; these may include degradative transfer to solvent and also primary radical termination. Bradbury and Melville (1954) showed that δ (in mole$^{1/2}$ l.$^{-1/2}$ sec$^{1/2}$) for *n*.butyl acrylate increases from 0·76 for bulk monomer to 3·5 for monomer at about 0·5 mole/l. in benzene; for styrene, δ was 43·5 for bulk monomer and 55·4 for monomer at 1·0 mole/l. The variation of δ with composition is anomalous since this quantity is a function of velocity constants only; it must mean that the derived value of δ is not really $k_t^{1/2}/k_p$ but actually includes contributions from the velocity constants for processes other than mutual termination and propagation. The apparent value of δ for a monomer may be affected by the presence of another monomer. In calculating ϕ, Bradbury and Melville made allowance for the effect of benzene upon δ but not for the effect of the other monomer. If [benzene] is fixed, raising the ratio [styrene]/[butyl acrylate] would be expected to cause a considerable increase in the value of δ for butyl acrylate, and to reduce δ for styrene to a smaller extent. Inspection of (5) shows that on this basis the value of ϕ should be greatest when [styrene]/[butyl acrylate] is large—as is observed (see Fig. 6.3).

Values quoted for velocity constants for cross-terminations in co-polymerizations and other liquid phase radical chain reactions must be treated with great caution. Diffusion-control of termination reactions may have some significance, but certainly (5), which is used for calculating ϕ, is not always valid. The occurrence of termination other than by the interaction of polymer radicals should be allowed for the introduction of additional terms into the denominator of (5). This would lead to a reduction of ϕ; the magnitudes of these additional terms would depend upon the composition of the feed, and so an apparent variation of $k_{t_{ab}}$ and ϕ with composition could result.

E. Primary Radical Termination

The interaction of polymer radicals with the primary radicals derived from the initiating system is usually ignored in kinetic analyses of radical polymerizations. It is accepted, however, that this type of termination is the main one in emulsion polymerizations; its importance in γ-ray initiated polymerizations has been discussed (Chapiro, Magat, Sebban and Wahl, 1955). It has been shown more recently (Bamford, Jenkins and Johnston, 1959*c*; Henrici-Olivé and Olivé, 1960) that it may also be of considerable significance in certain homogeneous polymerizations with thermal and photosensitizers, and that it may be responsible for deviations from "ideal" kinetics. There are indications that in the polymerization of vinyl benzoate interaction between the primary radicals and polymer radicals of the stabilized type (see Chapter 4, A.*4*) may be significant (Litt and Stannett, 1960).

Primary radical termination is a reaction involving radicals of different types and it may therefore be favoured over reactions of pairs of similar radicals. Bamford, Jenkins and Johnston (1959*c*) concluded that at 60°C the velocity constant for the interaction of a polystyrene radical with a 2-cyano-2-propyl radical may be sixty times that for the interaction of a pair of the polymer radicals.

If primary radical transfer is absent, three reactions are possible for a primary radical:

initiation	$R\cdot + M \longrightarrow R.M\cdot$	(10)
primary radical termination	$R\cdot + P\cdot \longrightarrow R.P$	(11)
recombination	$2R\cdot \longrightarrow R_2$	(12)

(11) and (12) are written as combinations, but disproportionations may be also possible. Unless [monomer] is very low or the primary radical is so unreactive that (10) is very slow, (12) is of very little significance. It is possible usually, therefore, to consider a competition between (10) and (11). The relative importance of (11) must be increased by

(*a*) reducing [monomer] since this affects the rate of (10);
(*b*) raising [initiator] since this increases the stationary concentration of polymer radicals and, therefore, also the rate of (11);
(*c*) reducing the temperature, because E_{10} must be greater than E_{11}.

If primary radical termination becomes significant, the rate of initiation must fall and the stationary value of [P·] is no longer independent of [monomer]; the order of the overall polymerization with respect to initiator falls below 0·5, and the order with respect to monomer rises above 1.

It is important to consider how, for a particular initiator, primary radical termination might depend upon the nature of the monomer. Replacement of styrene by a less reactive monomer, such as vinyl acetate, causes k_{10} to fall; at the same time, k_{11} rises because the polyvinyl acetate radical is the more reactive polymer radical. These changes in k_{10} and k_{11} would work in the same direction and accentuate primary radical termination for the case of vinyl acetate.

The rate of initiation measured by the initiator-fragment method is actually the sum of the rates of all processes by which initiator may become incorporated in polymer. Allowance can be made for those fragments which enter polymer as a result of transfer to initiator (see Chapter 5, E), and in certain cases it is possible to correct for primary radical termination by combination. These cases concern primary radicals which are themselves unstable, and which may dissociate into a smaller radical and a stable molecule, a typical example being the benzoyloxy radical. A polymerizing system can be considered as one containing a mixture of two scavengers, viz. the monomer molecules and the polymer radicals, the former being at a much higher concentration than the latter but having a much lower reactivity. If [monomer] is held constant, an increase in the steady value of [P·] must cause a greater proportion of the benzoyloxy radicals to be captured before they can dissociate. The ratio of benzoyloxy to phenyl end-groups in the polymer should, therefore, rise as [P·] is increased; this can be achieved either by raising [dibenzoyl peroxide] or by adding a second initiator to the system (Allen and Bevington, 1960*b*).

Consider the competition between the reactions

$$C_6H_5.CO.O\cdot \longrightarrow C_6H_5\cdot + CO_2 \tag{13}$$

$$C_6H_5.CO.O\cdot + M \longrightarrow C_6H_5.CO.O.M\cdot \tag{14}$$

$$C_6H_5.CO.O\cdot + P\cdot \longrightarrow C_6H_5.CO.O.P \tag{15}$$

The fraction of the benzoyloxy radicals which enter polymer is given by

$$x = \frac{k_{14}[B\cdot][M] + k_{15}[B\cdot][P\cdot]}{k_{13}[B\cdot] + k_{14}[B\cdot][M] + k_{15}[B\cdot][P\cdot]}$$

where [B·] and [P·] = stationary concentrations of benzoyloxy and polymer radicals respectively, so that

$$\frac{x}{1-x} = \frac{k_{14}[M] + k_{15}[P\cdot]}{k_{13}}$$

Putting $$[\mathrm{P}\cdot] = R_p/k_p[\mathrm{M}]$$

$$\frac{x}{1-x} = \frac{k_{14}[\mathrm{M}]}{k_{13}} + \frac{k_{15}R_p}{k_{13}k_p[\mathrm{M}]} \tag{16}$$

and for the benzoyloxy radicals

$$\frac{\text{rate of termination}}{\text{rate of initiation}} = \frac{k_{15}[\mathrm{P}\cdot]}{k_{14}[\mathrm{M}]} = \frac{k_{15}R_p}{k_{14}k_p[\mathrm{M}]^2} \tag{17}$$

This treatment assumes that transfer to initiator does not interfere; this is so if

$$\frac{\text{total rate of incorporation of initiator fragments in polymer}}{[\text{initiator}]}$$

is constant over the whole range studied.

By using dibenzoyl peroxide labelled at specific sites with carbon-14, $x/(1-x)$ can be determined by end-group analysis of the polymer (see Chapter 3, D.*1*), provided that all the phenyl radicals formed in (13) subsequently enter polymer either by an initiation process or by primary radical termination. According to (16), both $k_{15}/k_{13}\,k_p$ and k_{14}/k_{13} can be evaluated from a plot of $x/(1-x)$ against the rate of polymerization at a fixed value of [monomer]. For styrene at 60°C, these ratios have the values $1\cdot3\times10^4$ sec and $2\cdot5$ mole^{-1} l.$^{+1}$ respectively. (17) can be used to assess the importance of primary radical termination by the benzoyloxy radical. Results displayed in Fig. 6.4 show that primary radical termination in this system is important only if [monomer] is low, or the polymerization is very rapid, but then there might be self-heating which would reduce the importance of primary radical termination relative to that of initiation.

The small primary radical diffuses more easily than a large polymer radical, and so the relative importance of primary radicals in termination processes may increase when the normal mutual termination becomes diffusion-controlled. This must lead to an order with respect to initiator of less than $0\cdot5$; according to Hayden and Melville (1960), termination in the polymerization of methyl methacrylate becomes diffusion-controlled at about 10% conversion, and it is significant that at that point in the reaction the order with respect to initiator begins to fall towards $0\cdot25$.

Degradative transfer might be regarded as a special case of primary radical termination. Radicals which might otherwise initiate, engage in alternative reactions including reaction with polymer radicals.

Most of the growing radicals in emulsion polymerizations are deactivated by reaction with primary radicals. The interval between the entry of a second primary radical into a reaction site and the termination reaction, depends upon the size of the site and the values of k_p and k_t for the monomer. For particles of the size usually present during the emulsion polymerization of styrene, and using values of k_p and k_t derived from bulk and solution polymerizations, it has been shown that the average interval between the growth reactions for a polystyrene radical is appreciably greater than the average life of a second primary radical in a reaction site. The second primary radical, therefore, has little chance of reacting with monomer, and there is a high probability of it engaging in termination. The possibility of the velocity constant for the primary radical termination being greater than the normal value of k_t enhances this effect. It must be noted that during its sojourn in the aqueous phase,

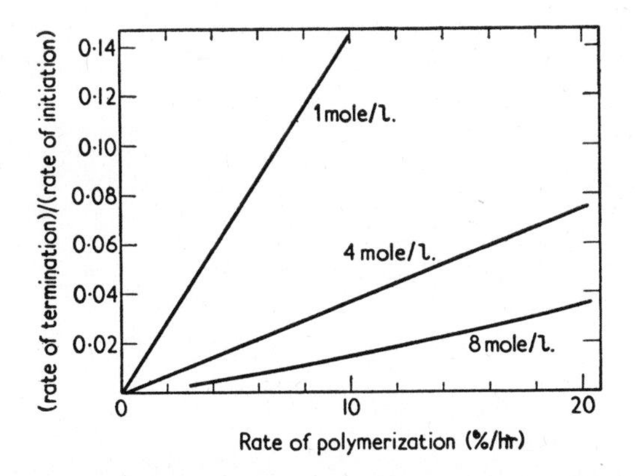

FIG. 6.4. (Rate of termination)/(rate of initiation) for benzoyloxy radicals in polymerization of styrene at 60°C for various values of [monomer] and rates of polymerization.

the primary radical may react with dissolved monomer. The radical which enters the reaction site, and which has been referred to as a primary radical, may therefore be larger than the radical originally generated.

Except in the very early stages of an emulsion polymerization, the rate is independent of [initiator] in the aqueous phase over a wide range. In homogeneous polymerizations, this result would imply that all termination is by primary radicals; the same is very nearly true for emulsion systems, but the zero-order relationship arises from a different reason. Primary radicals can be considered as entering the reaction sites at regular intervals; whether a site is active or inactive at a particular moment, depends whether an odd or an even number of radicals have entered since the start of the polymerization, provided that no radicals are lost from the site by diffusion into the surrounding aqueous phase. The average active and inactive periods are equal and may be a few seconds in duration. A reaction site is therefore, on the average, active for half the time; this is not affected by an alteration in the rate of formation of radicals in the aqueous phase, which can only change the frequency of alternation between the active and inactive periods of the particles.

CHAPTER 7

Retardation and Inhibition

A. CHARACTERISTICS OF RETARDATION AND INHIBITION

Inhibitors are substances which are considerably more reactive than monomers towards free radicals; when present in reaction mixtures even at low concentrations, they can compete successfully with the monomer for capturing the "available" primary radicals, although they do not interfere with geminate recombination of primary radicals (see Chapter 3, B). Retarders also can deactivate radicals but they are less reactive than inhibitors; at the concentrations ordinarily used, they do not prevent polymerization. If mutual termination does not occur and polymer radicals are destroyed singly by reaction with retarder, the rate of polymerization is proportional to rate of initiation instead of (rate of initiation)$^{1/2}$.

In all cases of retardation and inhibition, a reaction between the additive and a radical (primary or polymer) competes with one of the normal steps in the polymerization. The velocity constant for the reaction involving the additive depends upon the reactivity of that substance and also upon that of the radical. The magnitude of the effect of the additive upon the rate of polymerization, however, depends not only upon this velocity constant but also upon that of the normal reaction with which it competes; thus the effect of a particular additive may vary considerably from one monomer to another.

If primary radicals are generated in a mixture of an inhibitor and monomer, there is competition between the reactions

$$R\cdot + \text{inhibitor} \longrightarrow \text{inactive products} \quad (1)$$

$$R\cdot + M \longrightarrow R.M\cdot \quad (2)$$

Normally [monomer] is very much greater than [inhibitor], but k_1 is so much larger than k_2 that virtually all the radicals react according to (1). As the inhibitor is consumed, the balance between (1) and (2) gradually shifts; increasing numbers of the primary radicals are captured by monomer, and some of the inhibitor reacts with polymer radicals containing a small number of monomer units instead of with primary radicals. At a point depending upon the sensitivity of the method of measurement, the rate of consumption of monomer becomes significant and then gradually builds up to the value which would be observed in a system from which the inhibitor is absent; it may be necessary to make allowance for consumption of initiator during the inhibition period. The gradual increase in rate of polymerization is illustrated in Fig. 7.1, which refers to the polymerization of styrene in the presence of *p*.benzoquinone; this substance is rather an inefficient inhibitor for the polymerization, so that appreciable quantities of monomer are consumed before the quinone is used up and the polymerization attains a steady rate. If an inhibitor is very efficient, i.e. k_1 is very large, the transition from a

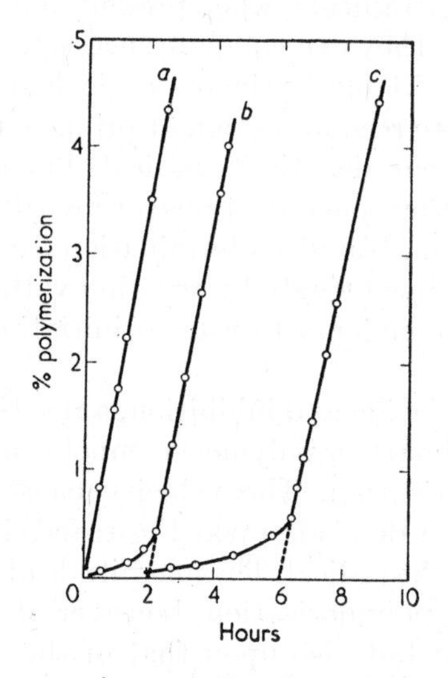

FIG. 7.1. Effect of *p*.benzoquinone upon polymerization of styrene at 60°C with [azo*iso*butyronitrile] = 0·5 g/l.; [quinone]—*a*, 0; *b*, 0·05; *c*, 0·123 g/l.

negligible rate of consumption of monomer to the full value is rather abrupt for all monomers.

The inhibition period is commonly quoted as the intercept on the time axis, as indicated in Fig. 7.1. Bamford, Jenkins and Johnston (1957) concluded that when an inhibitor is used to determine rates of initiation (see Chapter 3, C), the inhibition period should be taken as the time for the rate of polymerization to attain 64·8% of the final steady value. This conclusion refers to systems in which [monomer] and [initiator] do not change appreciably, and in which the products of (1) play no further part in the reaction. In many systems, however, these products act as retarders; consequently, as the initial concentration of inhibitor is raised, and the amounts of the derived retarders are increased, the rate of the subsequent polymerization becomes smaller.

Ideally an inhibitor should not become incorporated in high polymer since it should be converted into unreactive substances of low molecular weight. When ^{14}C-diphenylpicrylhydrazyl is used with styrene, however, the polymer formed in the later stages of the reaction contains combined radio-activity (Bevington, 1956).

If the rate of production of radicals from a sensitizer is almost unaffected by the nature of the medium, the length of the inhibition period for given concentrations of sensitizer and inhibitor should be independent of the nature of the monomer and its concentration. This ideal state of affairs is not always observed; when, for example, the stabilized free radical

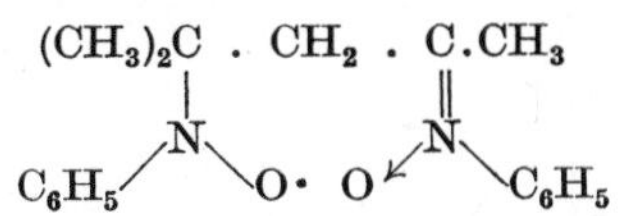

is used as an inhibitor for the polymerizations of styrene and methyl methacrylate with azo*iso*butyronitrile as sensitizer, there are considerable differences between the lengths of the inhibition periods found for the two monomers (Bevington and Ghanem, 1956). Evidently, during the inhibition period, there are reactions in addition to (1) and (2); in this particular case, complications arise from instability of the inhibitor itself. It must be concluded that inhibition is seldom as simple as in the ideal case; the difficulty of assigning a chemical equation to the interaction (1) is discussed in Section B of this Chapter.

Retarders are less reactive than inhibitors and they allow polymer radicals to grow to an appreciable size before deactivating them. It is necessary, however, to examine the possibility of such substances reducing the rate of initiation, and the competitions between (3) and (4) and between (5) and (6) must be considered:

$$R\cdot + M \longrightarrow R.M\cdot \quad (3)$$

$$R\cdot + \text{retarder} \longrightarrow \text{unreactive products} \quad (4)$$

$$P\cdot + M \longrightarrow P.M\cdot \quad (5)$$

$$P\cdot + \text{retarder} \longrightarrow \text{unreactive products} \quad (6)$$

If high polymer is produced although the additive is present, the rate of (5) must be considerably greater than that of (6); consequently the rate of (3) must be much larger than that of (4) unless k_4/k_3 is much greater than k_6/k_5. During the polymerization of methyl methacrylate with azo*iso*butyronitrile as photosensitizer, the rate of initiation is not much affected by the presence of the retarder *p*.benzoquinone; the scatter of results evident in Fig. 7.2 is largely associated with difficulty in reproducing the light intensity from one experiment to another. When, however, the N-oxide

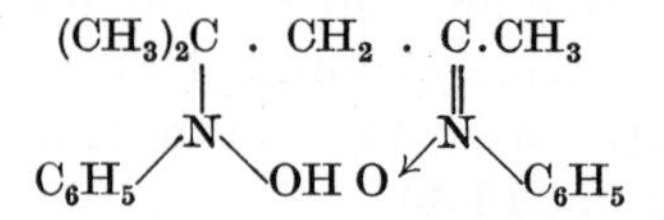

retards the polymerization of styrene, it also reduces the rate of initiation by radicals derived by thermal dissociation of the azo compound (see Fig. 7.2).

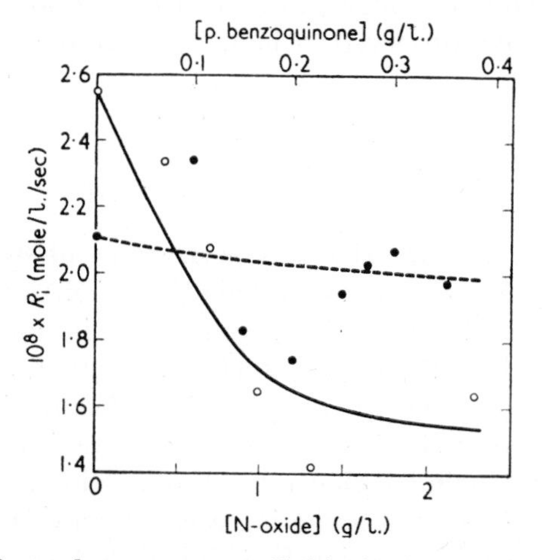

Fig. 7.2. Effects of retarders upon rates of initiation in sensitized polymerizations. —○— styrene at 60°C with azo*iso*butyronitrile at 0·30 g/l. and N-oxide; ——●—— methyl methacrylate at 25°C with fixed [azo*iso*butyronitrile] and light intensity, and *p*.benzoquinone.

The most useful classification of retarders and inhibitors is according to the way in which they interact with radicals; on this basis, there are three types of additive:

(*a*) *stabilized free radicals* which can react with radicals of the types present in polymerizing systems but not with monomers;
(*b*) substances which are essentially *transfer agents*; the products of the radical-displacements are, however, so unreactive that they re-initiate very inefficiently;
(*c*) substances to which primary or polymer radicals may become attached chemically, giving new radicals which are relatively unreactive; processes of this type can be considered as *co-polymerizations of a special type*.

Subsequent discussion will be based on this classification. No attempt will be made to discuss or even to list all the compounds which reduce the rates of radical polymerizations. In many cases, there has been only cursory examination of their effects, and it would be unwise to draw firm conclusions from the published results. In the following sections, some of the better-known retarders and inhibitors are discussed in some detail; experience gained from study of these can materially assist in the elucidation of the actions of the lesser known additives.

Evidence concerning the precise mechanism of the action of a retarder can be obtained from studies of several types, viz.

(*a*) the use of model radicals and identification of the products of their interaction with the retarder:

(*b*) kinetic studies of the retarded polymerization, in conjunction with measurements of the molecular weights of the resulting polymers;

(*c*) analysis of polymers for combined retarder, in particular determination of the number of retarder molecules incorporated in the average polymer molecule, and analysis of the reaction mixtures for products of low molecular weight derived from the retarder:

(*d*) physical and chemical examination of polymer, prepared in the presence of a retarder, to discover the nature of the bonds by which the retarder is combined in the polymer:

(*e*) study of the effects of modifying the structure of the retarder upon its reactivity; the modification may be substitution of deuterium for hydrogen at selected sites to see if the rate-determining step involves the breakage of a particular bond.

Examples of the applications of these methods will be given during discussion of particular retarders. Generally, (*a*) gives valuable information, but it is subject to those limitations already mentioned in connection with other uses of model radicals. Method (*b*) may lead to formal representation of the reactions, but very frequently more than one reaction scheme can be used to derive kinetics which approximate to those found in practice. Analysis of polymers for combined retarder may allow rejection of some of the reaction schemes which are satisfactory from the kinetic point of view. The analytical approach does not indicate in what form the retarder is combined in the polymer, nor whether it is present as an end-group or as a unit in the body of the polymer chain; this information may result from application of method (*d*).

B. Radical Inhibitors

For a free radical to be useful as an inhibitor, it must be stabilized sufficiently for it not to react with monomers and initiate polymerization, but it must react readily with radicals of the types encountered in polymerizing systems. Probably the best known radical inhibitors are the hydrazyls, of which diphenylpicrylhydrazyl is the most widely used. The formula

$$(C_6H_5)_2N.\underset{\bullet}{N}.C_6H_2(NO_2)_3$$

is usually written, but the crystals may contain solvent of crystallization (Lyons and Watson, 1955). The stability is due to delocalization of the unpaired electron and a number of resonance structures can be written. Electron spin resonance spectroscopy has shown that there is equal interaction between the unpaired electron and the two adjacent nitrogen atoms (Ingram, 1958). The hydrazyls are intensely coloured and can be determined at very low concentrations by spectrophotometry. The diphenylpicryl compound is used for measuring rates of production of radicals, both by dissociation of sensitizers and by other means; it is also used as a standard in electron spin resonance work.

It has been pointed out (see Section A) that diphenylpicrylhydrazyl is not an ideal inhibitor for some systems since products formed from it are reactive and may become incorporated in the polymer; the effect is most probably associated with the nitro groups. In the case of vinyl acetate (Bengough, 1955), the time required for disappearance of the colour of the hydrazyl is appreciably less than the time required for the rate of consumption of monomer to become significant; evidently, products formed from the hydrazyl can act as inhibitors. This discrepancy is not observed if vinyl acetate is replaced by methyl methacrylate; it is significant that polymethyl methacrylate radicals are less reactive than polyvinyl acetate radicals towards nitro compounds (Kice, 1954). Other radical inhibitors give less reactive products and those formed from

$$\begin{array}{ccccc} & (CH_3)_2C & . \; CH_2 \; . & C.CH_3 & \\ & | & & \| & \\ C_6H_5 - & N - O\cdot & & O \leftarrow N & - C_6H_5 \end{array}$$

for example, do not interfere in the polymerization of styrene (Bevington and Ghanem, 1956).

A peculiar result is obtained when diphenylpicrylhydrazyl is used during the thermal unsensitized polymerization of styrene (Russell and Tobolsky, 1953). The rate of consumption of hydrazyl is considerably greater than the rate of initiation calculated from the rate of polymerization observed in a parallel experiment without inhibitor. The discrepancy is believed not to be due to any abnormality in the behaviour of the inhibitor. It is thought that diradicals are formed at quite an appreciable rate, and that most of them are normally lost from the system by processes such as ring closure; the average life of these radicals is, however, sufficient for them to react with the hydrazyl. The rate of initiation of polymerization is low, because only a small proportion of the diradicals are converted by transfer reactions to mono-radicals which develop into long polymer chains.

The simplest mechanism for the reaction between the hydrazyl and a radical, would be direct addition to the nitrogen atom to give a tetra-substituted hydrazine; there is, however, no evidence that this reaction occurs. Diphenylpicrylhydrazyl reacts with many organic compounds to give quantitative yields of diphenylpicrylhydrazine, thus hydrazobenzene readily loses hydrogen to give azobenzene at room temperature (Braude, Brook and Linstead, 1954) and transfer of hydrogen from azo*iso*propane to the hydrazyl occurs so readily that this scavenger cannot be used to monitor the production of radicals by photo-dissociation of the azo compound (Laita and Macháček, 1959). It has been suggested that diphenylpicrylhydrazyl might undergo disproportionation with certain radicals to give the corresponding hydrazine; this cannot be a universal mechanism for its scavenging action since some of the radicals with which it reacts quite readily, e.g. benzoyloxy and phenyl, do not possess hydrogen atoms at suitable sites. Further, there are only small yields of diphenylpicrylhydrazine when azo*iso*butyronitrile is decomposed in the presence of the hydrazyl (Bevington, 1956). In this system, however, there may be other products containing N—H bonds since treatment of the final solution regenerates the colour characteristic of the hydrazyls (Bawn and Verdin, 1960). It is believed that the overall reaction can be represented as

$$(CH_3)_2C(CN)\cdot + (C_6H_5)_2N.\dot{N}.C_6H_2(NO_2)_3 \longrightarrow [(CH_3)_2C(CN){-}C_6H_4](C_6H_5)N{-}NH.C_6H_2(NO_2)_3 \quad (7)$$

but quite obviously this cannot occur in a single stage. This proposal that the radical may become attached to one of the phenyl groups of the hydrazyl had previously been made by Poirier, Kahler and Benington (1952). It also fits some observations made when polymethyl methacrylate is degraded in the presence of diphenylpicrylhydrazyl; some of the hydrazyl becomes chemically attached to the polymer (Henglein, 1955), but subsequent treatment of the polymer with lead dioxide regenerates the colour of the hydrazyl. It appears that the polymer radical may be attached to the scavenger as in (7).

Verdin (1960) reported that when azo*iso*butyronitrile is decomposed in the presence of diphenylpicrylhydrazyl and air, a substituted hydrazine is not formed. Evidently, the chemistry of the hydrazyl is further complicated by the presence of oxygen. It appears that the 2-cyano-2-propyl radicals formed from the azonitrile may react with oxygen instead of with the hydrazyl, and that the reaction of the resulting peroxy radicals, $(CH_3)_2C(CN).O.O\cdot$, with the scavenger cannot be represented by an overall equation similar to (7).

Many other stabilized free radicals have been studied by electron spin resonance spectroscopy, and a list has been compiled (Ingram, 1958). For only a few of these radicals have there been detailed chemical studies; it seems to be impossible to construct a single simple chemical equation which represents the interaction of radicals of all types with a scavenger. All the stabilized radicals owe their stability to delocalization of the unpaired electron; the electron density at several positions may be sufficient for reaction with other radicals, and the characteristics of a radical may determine at which site in the scavenger it reacts.

C. Retardation by Transfer Agents

When the radical formed in the first stage of a transfer reaction is unreactive towards monomer, the efficiency of re-initiation is low, and transfer is accompanied by a reduction in the overall rate of consumption of monomer (see Chapter 5, G). Any transfer agent showing this effect could be regarded as a retarder, but the description is usually reserved for cases in which efficiency of re-initiation approaches zero.

Compounds of the general formula X.H, where X· represents a highly stabilized radical of the type considered in the previous section, are of interest as potential retarders for polymerizations. The X—H bond is comparatively weak, so that the hydrogen-abstraction

$$P\cdot + X.H \longrightarrow P.H + X\cdot \tag{8}$$

occurs readily, and the radical X· is so unreactive that it cannot initiate polymerization. Generally, it is expected that pairs of X· radicals will not interact and that they will be consumed in the reaction

$$P\cdot + X\cdot \longrightarrow \text{unreactive products} \tag{9}$$

Thus one molecule of the retarder stops the growth of two chains; if [XH] is sufficient to suppress mutual termination of polymer radicals, and if (9) occurs by combination, the average number of retarder molecules combined in a polymer molecule is 0·5. In many cases, however, there must be considerable uncertainty about the exact mechanism of (9), as explained in Section B of this Chapter.

One of the retarders of this class is diphenylpicrylhydrazine, X in (9) representing the stabilized radical diphenylpicrylhydrazyl. The hydrazine is a retarder for the sensitized polymerization of styrene at 60°C, and it becomes chemically incorporated in the polymer (Bevington and Ghanem, 1958). Experiments with labelled initiator and labelled retarder showed that, if [retarder] is high enough, the numbers of initiator fragments and retarder molecules per polymer molecule are close to 1 and 0·5, respectively, as required by the general reaction scheme of (8) followed by (9). The hydrazine has been used (Laita and Macháček, 1959) to measure the rate of production of radicals by photo-dissociation of azo*iso*propane, one molecule of the hydrazine being taken as equivalent to two radicals. Verdin (1960) showed that although 2-cyano-2-propyl radicals show little tendency to abstract hydrogen from diphenylpicrylhydrazine, the peroxy radical $(CH_3)_2C(CN).O.O\cdot$ (made by decomposing azo*iso*butyronitrile in the presence of oxygen) engages in this reaction quite readily. The hydrazyl is formed and reacts with other peroxy radicals, but the mechanisms of these processes are not known. A point to be noted in connection with this substituted hydrazine is that it contains nitro groups; aromatic nitro compounds in general act as retarders, and so there may be two distinct ways in which this particular compound may interfere in polymerizations.

Another retarder to be considered is

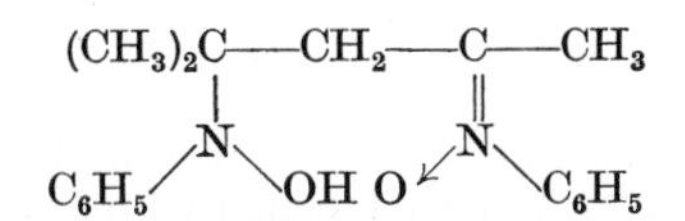

This N-oxide is related to one of the stabilized radicals mentioned in Section B. It retards the sensitized polymerization of styrene at 60°C (Bevington and Ghanem, 1958) (see Fig. 7.3). The numbers of initiator fragments and retarder molecules per polymer molecule approach 1·0 and 0·5, respectively, as [retarder] increases, so satisfying the reaction scheme given earlier in this section. This retarder might be used for measurements of rates of initiation by an analytical technique (see Chaption 3, C), each molecule of retarder included in the polymer being equivalent to two growing chains. Three pieces of evidence indicate that this procedure is not a general one and that it may be unreliable:

(*a*) as shown in Fig. 7.2, the N-oxide may react preferentially with primary radicals of certain types; thus, the rate of initiation derived from analysis of polymers prepared in the presence of the retarder may be less than the rate of production of radicals capable of initiating polymerization in the absence of retarder;

(*b*) the N-oxide does not retard the polymerization of methyl methacrylate;

(*c*) presence of the compound during the decomposition of azo*iso*butyronitrile in benzene affects the yields of the various products formed from the 2-cyano-2-propyl radicals; if the first reaction is according to (8), substantial quantities of *iso*butyronitrile should be produced, but in practice this is not so, and, therefore, it is doubtful whether the general equations (8) and (9) are adequate.

Triphenylmethane is a very powerful retarder for the polymerization of vinyl acetate but has hardly any effect upon the rate of polymerization of styrene. For quite low values of [triphenylmethane] in vinyl acetate,

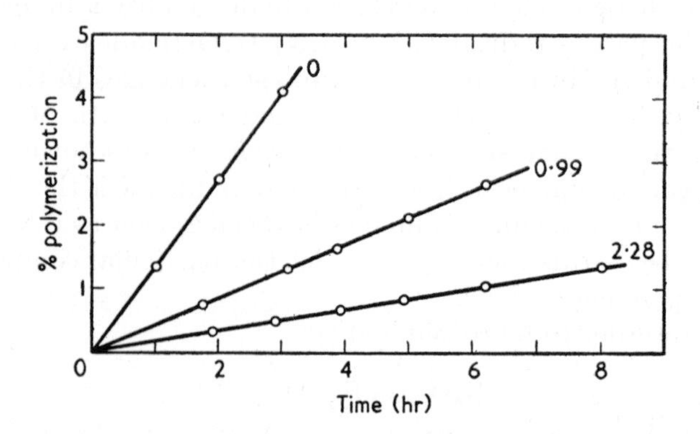

FIG. 7.3. Effects of a N-oxide upon the polymerization of styrene at 60°C, [azo*iso*butyronitrile] = 0·30 g/l., lines labelled with [retarder] in g/l.

the polymerization is first order with respect to initiator, indicating that for this monomer the efficiency of re-initiation by the triphenylmethyl radical is almost zero (see Chapter 5, G). The high reactivity of the polyvinyl acetate radical means that k_8 is large, and the corresponding low reactivity of monomeric vinyl acetate means that the rate of re-initiation is small; these together make the polymerization of vinyl acetate particularly sensitive to retardation, not only by triphenylmethane, but also by other substances which can react by the same general mechanism. These factors are responsible also for the even greater effects of impurities upon the polymerization of vinylene carbonate (Smets and Hayashi, 1958).

Certain aromatic secondary amines, e.g. diphenylamine, are used as anti-oxidants, so that evidently they can interfere in radical chain reactions. These substances have not been much studied in polymerizing systems, but diphenylamine is quite a powerful retarder for the polymerization of vinyl acetate at 60°C (Bevington and Troth, 1960), although it has little effect on certain other polymerizations (Kice, 1954; Angert and Kuzminskiĭ, 1958). It is tempting to suppose that the first stage in any retardation would be

$$P\cdot + (C_6H_5)_2NH \longrightarrow P.H + (C_6H_5)_2N\cdot \tag{10}$$

to give the unreactive diphenylnitrogen radical which would then terminate another growing chain. Boozer, Hammond, Hamilton and Sen (1955) believe that this reaction does not occur, basing this view on their observation that $(C_6H_5)_2NH$ and $(C_6H_5)_2ND$ have identical effects upon oxidations of hydrocarbons; if (10) were the first step, its rate should be reduced by the introduction of deuterium into the molecule. They pointed out also that some powerful amine anti-oxidants, such as

have no labile hydrogen atoms. They supposed that the first stage is the formation of a loose molecular complex of peroxy radical with amine, followed by further unspecified reaction with a peroxy radical to give unreactive products. Angert and Kuzminskiĭ (1958) concluded, however, from a study of the effect of phenyl-β-naphthylamine upon the oxidation of rubber, that the first stage is according to (10), with the radical P• being one in which the unpaired electron is on the oxygen atom. The chief evidence in support of this belief is that the inhibitory power of an amine for the oxidation of rubber can be correlated with the ease with which the N—H bond can be broken, this being assessed by the readiness with which diphenylpicrylhydrazyl abstracts hydrogen from the secondary amine. Shelton, McDonel and Crano (1960) agree that diphenylamine reacts according to (10) and have accounted for the previous failure to detect an isotope effect; they showed that deuterium can be lost from the amine very readily by an exchange process with traces of water in the reaction mixture.

Similar problems of mechanism arise in connection with the use of certain substituted phenols as inhibitors of oxidations. The stabilized radical 2,4,6-triphenylphenoxy has been described as an inhibitor for radical polymerizations (Breitenbach, Olaf and Schindler, 1957); other similar radicals are probably unreactive and inefficient initiators of polymerizations, even if not stable enough to be isolated. The corresponding phenols (X.OH) might be retarders acting according to the scheme

$$P\cdot + X.O.H \longrightarrow P.H + X.O\cdot \tag{11}$$

$$X.O\cdot + P\cdot \longrightarrow X.O.P \tag{12}$$

The existence of a relationship between the oxidation-reduction potentials of a series of phenolic compounds and their inhibitory powers during the oxidation of ethyl linoleate is significant (Bolland and ten Have, 1947); it suggests that the inhibition process involves loss of the hydrogen atom from the anti-oxidant. The effects of phenols upon radical polymerizations have received little attention; if they could cause retardation by (11) and (12), the following effects would be expected:

(*a*) replacement of XOH by XOD should reduce the inhibitory power of the phenol since the rate of (11) would be reduced by an isotope effect;

(*b*) if [phenol] is high enough to ensure that all reaction chains are terminated by (11) and (12), the average polymer molecule should contain 1 initiator fragment and 0·5 molecule of retarder:

(*c*) the radical X· would be attached to the polymer through an ether linkage which could be broken, but as it forms an end-group its removal would hardly affect the average molecular weight of the polymer.

According to Boozer, Hammond, Hamilton and Sen (1955), alkylperoxy radicals, as found in oxidizing systems, may react with phenolic compounds by a two-stage process in which (11) is followed by

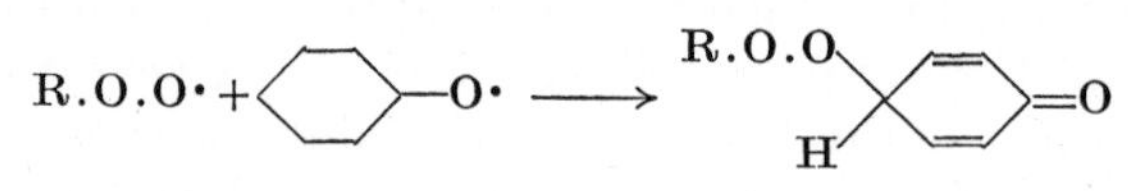

instead of by (12).

Ferric chloride reacts with polymer radicals by an electron-transfer process; the reaction has been mentioned in connection with measurement of rates of initiation (see Chapter 3, C) and also as an example of a process in which the rate of reaction is strongly dependent upon polar factors (see Chapter 5, B). The reaction can be formulated in two ways

$$\mathrm{P.CH_2.CHX\cdot + FeCl_3} \begin{cases} \nearrow \mathrm{P.CH_2.CHXCl + FeCl_2} \\ \searrow \mathrm{P.CH{:}CHX + HCl + FeCl_2} \end{cases} \qquad (13)$$

The relative importances of these courses may vary from one system to another. They are written as involving ferric chloride, but actually a complex with dimethyl formamide, used as solvent, may be involved (Bamford, Jenkins and Johnston, 1957); thus, for polyacrylonitrile radicals at 60°C, dimethylformamide at 0·06 mole/l. causes k_{13} to fall to about one-fortieth of the value found when the solvent is absent. (13) differs from radical-displacement reactions in that it causes complete destruction of the radical.

If [ferric chloride] is high enough, mutual termination of polymer radicals is suppressed. If the rate of (13) is very much greater than that of propagation, inhibition is observed; this is so for the case of styrene, the rate of consumption of monomer being very low until the ferric chloride is consumed. With acrylonitrile and methacrylonitrile, only retardation is observed; for the polymer radicals corresponding to these monomers, the values of k_p/k_{13} are large enough for propagation to occur to an appreciable extent.

TABLE 7.1

VELOCITY CONSTANTS FOR REACTIONS OF POLYMER RADICALS WITH FERRIC CHLORIDE AT 60°C IN DIMETHYLFORMAMIDE

Polymer radical	k_p/k_{13}	k_{13} (mole^{-1} l.$^{+1}$ sec^{-1})
styrene	$1\cdot9\times10^{-3}$	$5\cdot4\times10^{4}$
acrylonitrile	$0\cdot30$	$6\cdot5\times10^{3}$
methacrylonitrile	$0\cdot32$	$6\cdot1\times10^{2}$

In those systems in which retardation is observed, the polymerization is of order $+1$ with respect to initiator and of order -1 with respect to the retarder over a wide range of values of [ferric chloride], showing that all reaction chains are terminated by (13); over this range, the rate of production of ferrous chloride in the system is independent of [ferric chloride] and directly proportional to [initiator]. This shows that measurements of the rate of formation of the ferrous ion can legitimately be used to determine rates of initiation. It was deduced that normally less than 10% of the primary radicals react directly with the retarder.

There may be quite large differences between the effects of various ferric salts upon a polymerization (Entwistle, 1960). With methyl methacrylate, the salicylate is a weak retarder, the benzoate a more powerful retarder, and the bromide an inhibitor. As might be expected, the velocity constant for

$$\mathrm{P.CH_2.CHX\cdot + FeY_3 \longrightarrow P.CH_2.CHXY + FeY_2}$$
$$\text{or} \longrightarrow \mathrm{P.CH{:}CHX + HY + FeY_2}$$

depends critically upon the nature of Y as well as upon that of the polymer radical.

D. Retardation by Co-polymerization

The rate at which vinyl acetate polymerizes can be reduced very considerably by the presence of small quantities of a more reactive monomer, such as styrene, which gives rise to a polymer radical less reactive than the polyvinyl acetate radical. Because of the favouring of cross-propagation, the reactive polyvinyl acetate radicals are converted to less reactive polymer radicals and this produces a reduction in the overall rate of polymerization. Of the many examples of systems in which this effect is found, some of the most interesting involve a second monomer which does not contain an olefinic double bond. Just as in the case of retardation by transfer, retardation by co-polymerization is most likely to occur during the polymerization of a monomer, such as vinyl acetate, which, though relatively unreactive itself, gives rise to a very reactive polymer radical.

Free radicals are generally quite reactive towards oxygen, and there are many examples of the addition reaction

$$R\cdot + O_2 \longrightarrow R.O.O\cdot$$

If a polymer radical reacts in this way, the oxygen can be regarded as a co-monomer (see Chapter 4, B.*1*). The magnitude of its effect upon the overall rate of polymerization depends upon the values of the velocity constants for the cross-propagations

$$P\cdot + O_2 \longrightarrow P.O.O\cdot \quad (14)$$

$$P.O.O\cdot + M \longrightarrow P.O.O.M\cdot \quad (15)$$

In many cases, k_{14} is quite large and k_{15} very small, so that the oxygen acts as a retarder; this is so for methyl methacrylate (Schulz and Henrici, 1956). The transition from almost complete inhibition to the full rate of polymerization as [oxygen] becomes very small, may be quite abrupt. Low co-polymers of some monomers with oxygen have been isolated (Barnes, Elofson and Jones, 1950) and shown to contain peroxidic groups as required by the reaction scheme described above. For some monomers, co-polymerization with oxygen may be quite rapid; these monomers are ones like α- and β-methyl styrene, which normally polymerize very slowly by radical mechanisms. 1:1 co-polymers are produced and the systems resemble co-polymerizations of monomers, such as maleic anhydride and stilbene, neither of which polymerizes on its own (Mayo, Miller and Russell, 1958). It was concluded that the reactivity of the peroxy polymer radical, P . O . O·, is affected to some extent by the nature of the radical P·, so providing another example of the effect of a penultimate group upon the reactivity of a polymer radical.

The polyperoxide formed by co-polymerization of a monomer with oxygen may decompose to radicals capable of initiating polymerization; this is most likely to happen at fairly high temperatures. Oxygen completely inhibits thermal polymerizations of methacrylonitrile but is less effective in photo-polymerizations; this can be attributed to photolysis of the peroxidic groups of the low co-polymer of the monomer and oxygen (Strause and Dyer, 1956). Under some conditions, therefore, oxygen can exert a delayed accelerating effect on polymerizations; a period of inhibition or pronounced retardation can be followed by a reaction proceeding rather quickly. Each oxygen unit in the co-polymer can give rise to two radicals, so that the case might be regarded as one of degenerate kinetic branching.

In many ways, the effects of sulphur on vinyl polymerizations resemble those of oxygen. It can act as an inhibitor or a strong retarder by the co-polymerization mechanism; radicals $P.(S)_x\cdot$ thus formed, are not efficient for further propagation of the reaction chains and tend to engage in termination reactions (Bartlett and Trifan, 1956). Sulphur-containing radicals of this type may be produced also in transfer reactions and cause retardation by failing to re-initiate efficiently (see Chapter 5, E). Co-polymers containing sulphur units may initiate polymerization at high temperatures, reactive radicals being formed by scission of S—S bonds; thus, although sulphur reduces the rate of polymerization of styrene in the early stages of the reaction at 80°C, the rate in the later stages is higher than expected.

The quinones are important examples of retarders of the co-monomer type. The magnitude of the effect exerted upon a polymerization by a quinone depends not only upon its own nature but also upon that of the monomer; this can be illustrated by comparing the effects of *p*.benzoquinone upon the polymerizations of styrene and methyl methacrylate (Bevington, Ghanem and Melville, 1955*a*, *b*). The quinone reduces the rate of polymerization of styrene very considerably; the additive is completely consumed in the early stages of the reaction and it can be regarded as an inhibitor, although not a very effective one (see Fig. 7.1). On the other hand, *p*.benzoquinone introduces no inhibition period into the polymerization of methyl methacrylate (see Fig. 7.4); although retardation is pronounced, the rate of consumption of monomer is steady over a considerable part of the reaction.

In the early stages of the retarded polymerization of styrene, only low polymer is produced; it contains combined *p*.benzoquinone in quantities which may be equivalent to more than one quinone unit per polymer molecule. A very large part of the combined quinone can be removed from the polymer by treatment with a reagent which cleaves ether linkages, and this removal is accompanied by a substantial reduction in the average molecular weight of the polymer (Bevington, Ghanem and Melville, 1955*b*).

Polymers of methyl methacrylate prepared in the presence of

p.benzoquinone also contain combined retarder. As [*p*.benzoquinone] is increased, the order of the polymerization with respect to initiator rises towards 1; at the same time, the number of molecules of quinone combined in each polymer molecule becomes close to 1 and the corresponding number of initiator fragments is 2 (Bevington, Ghanem and Melville, 1955*a*). It appears that the first stage in the retardation for both styrene and methyl methacrylate is

$$\text{P}\cdot + \text{O}{=}\langle\overline{\quad}\rangle{=}\text{O} \longrightarrow \text{P—O—}\langle\overline{\quad}\rangle\text{—O}\cdot \qquad (16)$$

and that this reaction occurs rather more readily, relative to the normal propagation, for the polystyrene radical than for the polymethyl

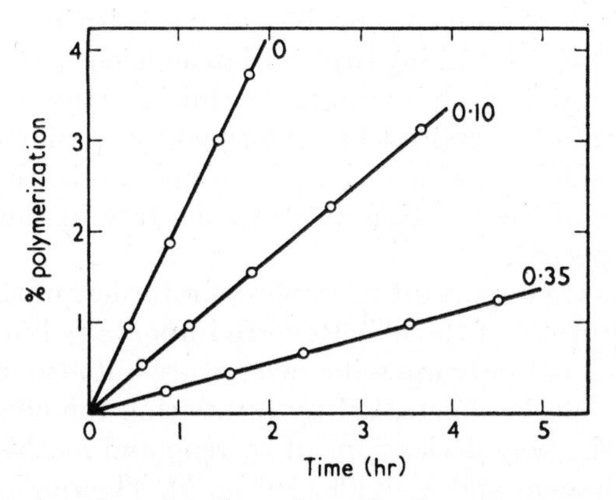

FIG. 7.4. Conversion-time plots for polymerization of bulk methyl methacrylate at 25°C using fixed light source and [azo*iso*butyronitrile]; lines labelled with initial value of [*p*.benzoquinone] in g/l.

methacrylate radical. The product of (16) is rather unreactive; in the case of methyl methacrylate, it shows little tendency to react with monomer and is consumed in the termination

$$\text{P—O—}\langle\overline{\quad}\rangle\text{—O}\cdot + \text{P}\cdot \longrightarrow \text{P—O—}\langle\overline{\quad}\rangle\text{—O—P}$$

but in the case of styrene, the growth reaction

$$\text{P—O—}\langle\overline{\quad}\rangle\text{—O}\cdot + \text{CH}_2\text{:CH.C}_6\text{H}_5 \longrightarrow \text{P—O—}\langle\overline{\quad}\rangle\text{—O—CH}_2\text{.CH(C}_6\text{H}_5)\cdot \qquad (17)$$

is significant, so that more than one molecule of quinone may be included in each polymer molecule. Tüdös (1958) has deduced that E_{16} is about 4 kcal/mole and E_{17} about 9 kcal/mole. For certain other quinones, e.g.

chloranil, the co-polymerization of retarder and monomer is more pronounced; this applies to both styrene (Breitenbach and Tschamler, 1951) and methyl methacrylate (Kice, 1954).

Other evidence that radicals attack *p*.benzoquinone and chloranil at the oxygen atom has been gathered from studies with small radicals (Bickel and Waters, 1950*b*). These authors referred to earlier workers' conclusions that nuclearly substituted quinonoid products are formed by attack of radicals on quinones, but their own work with $(CH_3)_2C(CN)\cdot$ and $(CH_3)_2C(COOCH_3)\cdot$ showed that all products could be accounted for by supposing that the first reaction is similar to (16).

Study of the effects of *p*.benzoquinone upon the rate of polymerization of methyl methacrylate, and upon the average molecular weight of the product, is significant in connection with the mechanism of mutual termination in the polymerization of this monomer (Bonsall, Valentine and Melville, 1953). It was found that the reduction in the molecular weight of the polymer resulting from the presence of the retarder was not as pronounced as the reduction in the rate of polymerization. This effect would be found if in the relationship

$$\text{average degree of polymerization} = x(\text{kinetic chain length})$$

where $1 \leqslant x \leqslant 2$ if transfer is negligible, the value of x in the retarded polymerization is greater than in the unretarded reaction. It was deduced that the mutual termination

$$2P\cdot \longrightarrow \text{unreactive products}$$

is by disproportionation, while the termination in the retarded polymerization

$$P\cdot + P.Q\cdot \longrightarrow \text{unreactive products} \qquad (18)$$

occurs by combination. In (18), Q represents a benzoquinone unit in the polymer chain, but these studies give no information on the way in which it is linked to the polymer radical P·. Funt and Williams (1960) also have studied the effects of *p*.benzoquinone upon the polymerization of methyl methacrylate by analysing the resulting polymers for combined initiator and retarder. Their conclusions are somewhat different from those just presented; in particular, they attach very great significance to the interaction of primary radicals with the quinone.

Szwarc (1955) has compared the velocity constants for the reactions of quinones with the methyl radical by the technique used for comparing the reactivities of monomers towards this reference radical (see Chapter 3, D.*2*). Results collected in Table 7.2 illustrate the considerable variations between the reactivities of the various quinones. Kice (1954) showed that the velocity constant for the reaction at 44°C of a polymethyl methacrylate radical with *p*.benzoquinone is about twenty times that for the corresponding reaction involving the tetrachloroquinone (chloranil).

Aromatic nitro compounds form another important group of retarders; initiator and monomers containing aromatic nitro groups usually show anomalous behaviour during polymerization, attributable to effects of these groups. Retarders of this kind function by the co-polymerization mechanism, the polymer radical attaching itself to the molecule of retarder to give a less reactive radical which tends to engage in termination rather than in further growth reactions. The reactivity of a nitro compound as a retarder depends, of course, upon its structure, and

TABLE 7.2

RELATIVE REACTIVITIES OF QUINONES TOWARDS THE METHYL RADICAL

Substituent in *p*.benzoquinone	Relative reactivity
none	1·00
2-methyl	0·68
2-methoxy	0·52
2-chloro	1·72
2,5-dimethyl	0·43
2,5-dichloro	2·60
2,6-dichloro	2·54
tetrachloro	0·02

generally the reactivity increases sharply with the number of nitro groups. When polynitro compounds are used, it is usually possible to divide the polymerization into phases, the original retarder being converted into a less reactive retarder which contains fewer unreacted nitro groups, and which may form part of a polymer molecule (Bartlett and Kwart, 1952). The polymethyl methacrylate and polymethyl acrylate radicals are comparatively unreactive towards nitro compounds (Kice, 1954, 1956), most probably because these compounds function as electron acceptors.

The problem of the mechanism by which nitro compounds act as retarders has not yet been completely elucidated; Walling (1957) has summarized some of the ideas on the subject. Some workers have considered that the first step involves addition of the radical to the aromatic nucleus

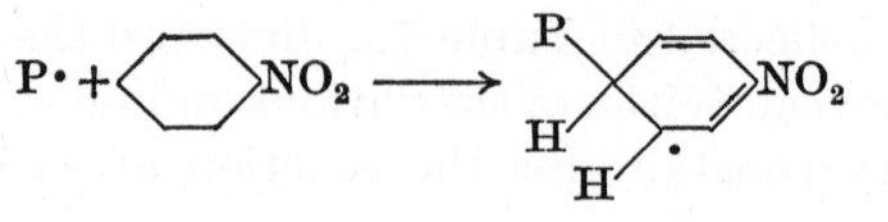

and others, for example Inamoto and Simamura (1958), that attack occurs on the nitro group itself:

$$\mathrm{P\cdot + O_2N{-}C_6H_5 \longrightarrow \begin{matrix} P{-}O \\ \cdot O \end{matrix}\!\!>\!N{-}C_6H_5} \qquad (19)$$

Experimental evidence can be selected to support both views, and it is likely that no single mechanism is applicable to all nitro compounds and all radicals.

Samples of polystyrene prepared at 60°C, with azo*iso*butyronitrile as initiator and *m*.dinitrobenzene as retarder, were analysed for combined

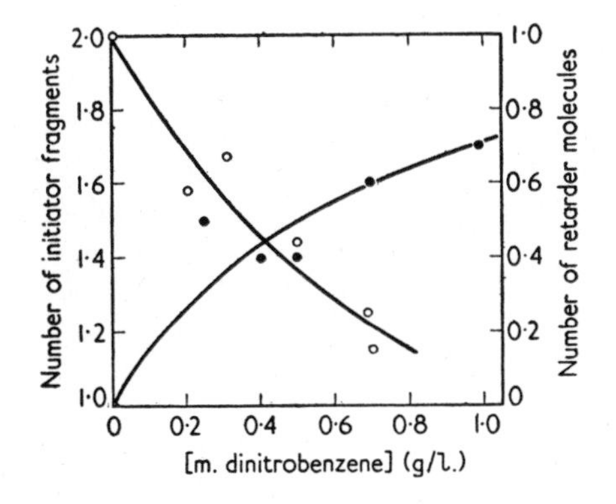

FIG. 7.5. Numbers of initiator fragments, ○, and retarder molecules, ●, per molecule of polystyrene. Polymers prepared at 60°C with [azo*iso*butyronitrile] = 0·30 g/l. and various concentrations of *m*.dinitrobenzene.

initiator and retarder (Bevington and Ghanem, 1959) (see Fig. 7.5). The results fit a reaction scheme in which (19) is usually followed by

$$\mathrm{C_6H_5{-}N\!\!<\!\begin{matrix} O.P \\ O\cdot \end{matrix} + P.CH_2.CH(C_6H_5)\cdot \longrightarrow C_6H_5{-}N\!\!<\!\begin{matrix} O.P \\ O.H \end{matrix} + P.CH{:}CH.C_6H_5} \qquad (20)$$

but the reaction

$$\mathrm{C_6H_5{-}N\!\!<\!\begin{matrix} O.P \\ O\cdot \end{matrix} + CH_2{:}CH.C_6H_5 \longrightarrow C_6H_5{-}N\!\!<\!\begin{matrix} O.P \\ O.CH_2.CH(C_6H_5)\cdot \end{matrix}} \qquad (21)$$

occurs occasionally. If (20) is the sole termination step and transfer reactions are absent, each polymer molecule contains one initiator fragment corresponding to the end-group of P·. In the absence of (21), the average number of retarder molecules per polymer molecule would be 0·5; the fact that the number can rise to about 0·7 indicates that, on the average, the polymer consists of two molecules containing no combined retarder to one containing two molecules of dinitrobenzene. It was shown that a large part of the retarder could subsequently be removed from the polymer by treatment with an ether-cleaving reagent. There is a formal similarity between the system and the one involving *p*.benzoquinone.

References

Alfrey, T., Bohrer, J. J. and Mark, H. (1952)."Co-Polymerization". Interscience Publishers Inc., New York.
Alfrey, T. and Price, C. C. (1947). *J. Polym. Sci.* **2,** 101.
Allen, J. K. and Bevington, J. C. (1960*a*). Not yet published.
Allen, J. K. and Bevington, J. C. (1960*b*). *Trans. Faraday Soc.* **56,** 1762.
Allen, P. W., Ayrey, G. and Moore, C. G. (1959). *J. Polym. Sci.* **36,** 55.
Allen, P. W. and McSweeney, G. P. (1958). *Trans. Faraday Soc.* **54,** 715.
Allen, P. W. and Merrett, F. M. (1956). *J. Polym. Sci.* **22,** 193.
Allen, P. W., Merrett, F. M. and Scanlan, J. (1955). *Trans. Faraday Soc.* **51,** 95.
Angert, L. G. and Kuzminskiĭ, A. S. (1958). *J. Polym. Sci.* **32,** 1.
Arlman, E. J. (1955). *J. Polym. Sci.* **17,** 375.
Atherton, N. M., Melville, H. W. and Whiffen, D. H. (1959). *J. Polym. Sci.* **34,** 199.
Ayrey, G. and Moore, C. G. (1959). *J. Polym.* Sci. **36,** 41.
Bailey, B. E. and Jenkins, A. D. (1960). *Trans. Faraday Soc.* **56,** 903.
Bailey, H. C. and Godin, G. W. (1956). *Trans. Faraday Soc.* **52,** 68.
Baker, C. A. and Williams, R. J. P. (1956). *J. chem. Soc.* 2352.
Bamford, C. H., Barb, W. G., Jenkins, A. D. and Onyon, P. F. (1958). "The Kinetics of Vinyl Polymerizations by Radical Mechanisms". Butterworth & Co. (Publishers) Ltd., London.
Bamford, C. H. and Jenkins, A. D. (1955) *Nature, Lond.* **176,** 78.
Bamford, C. H. and Jenkins, A. D. (1960). *Trans. Faraday Soc.* **56,** 907.
Bamford, C. H., Jenkins, A. D. and Johnston, R. (1957). *Proc. roy. Soc.* A, **239,** 214.
Bamford, C. H., Jenkins, A. D. and Johnston, R. (1958). *J. Polym. Sci.* **29,** 355.
Bamford, C. H., Jenkins, A. D. and Johnston, R. (1959*a*). *Trans. Faraday Soc.* **55,** 179.
Bamford, C. H., Jenkins, A. D. and Johnston, R. (1959*b*). *Trans. Faraday Soc.* **55,** 418.
Bamford, C. H., Jenkins, A. D. and Johnston, R. (1959*c*). *Trans. Faraday Soc.* **55,** 1451.
Bamford, C. H., Jenkins, A. D. and Wayne, R. P. (1960). *Trans. Faraday Soc.* **56,** 932.
Bamford, C. H. and White, E. F. T. (1956). *Trans. Faraday Soc.* **52,** 716.
Bamford, C. H. and White, E. F. T. (1958). *Trans. Faraday Soc.* **54,** 268.
Bamford, C. H. and White, E. F. T. (1960). *J. chem. Soc.* 4490.
Barb, W. G. (1953*a*). *J. Polym. Sci.* **10,** 49.
Barb, W. G. (1953*b*). *J. Polym. Sci.* **11,** 117.
Barb, W. G. (1953*c*). *J. Amer. chem. Soc.* **75,** 224.
Barb, W. G., Baxendale, J. H., George, P. and Hargrave, K. R. (1951). *Trans. Faraday Soc.* **47,** 462, 591.
Barnes, C. E., Elofson, R. M. and Jones, G. D. (1950). *J. Amer. chem. Soc.* **72,** 210.
Barson, C. A. and Bevington, J. C. (1958). *Tetrahedron,* **4,** 147.
Barson, C. A. Bevington, J. C. and Eaves, D. E. (1958). *Trans. Faraday Soc.* **54,** 1678.
Bartlett, P. D. and Altshul, R. (1945). *J. Amer. chem. Soc.* **67,** 812, 816.
Bartlett, P. D. and Hiatt, R. R. (1958), *J. Amer. chem. Soc.* **80,** 1398.

Bartlett, P. D. and Kwart, H. (1952). *J. Amer. chem. Soc.* **74,** 3969.
Bartlett, P. D. and Leffler, J. E. (1950). *J. Amer. chem. Soc.* **72,** 3030.
Bartlett, P. D. and Tate, F. A. (1953). *J. Amer. chem. Soc.* **75,** 91.
Bartlett, P. D. and Trifan, D. S. (1956), *J. Polym. Sci.* **20,** 457.
Basu, S., Sen, J. N. and Palit, S. R. (1952). *Proc. roy. Soc.* A, **214,** 247.
Bawn, C. E. H. and Margerison, D. (1955), *Trans. Faraday Soc.* **51,** 925.
Bawn, C. E. H. and Mellish, S. F. (1951). *Trans. Faraday Soc.* **47,** 1216.
Bawn, C. E. H. and Verdin, D. (1960). *Trans. Faraday Soc.* **56,** 815.
Bawn, C. E. H. and White, A. G. (1951). *J. chem. Soc.* 331, 339, 343.
Baxendale, J. H., Evans, M. G. and Park, G. S. (1946). *Trans. Faraday Soc.* **42,** 155.
Baysal, B. and Tobolsky, A. V. (1952). *J. Polym. Sci.* **8,** 529.
Bell, E. R., Raley, J. H., Rust, F. F., Seubold, F. H. and Vaughan, W. E. (1951). *Disc. Faraday Soc.* **10,** 242.
Bengough, W. I. (1955). *Chem. & Ind.* 599.
Bengough, W. I. and Melville, H. W. (1955). *Proc. roy. Soc.* A, **230,** 429.
Bengough, W. I. and Thomson, R. A. M. (1960). *Trans. Faraday Soc.* **56,** 407.
Benson, S. W. (1960). "The Foundations of Chemical Kinetics". McGraw-Hill Book Co. Inc., New York.
Benson, S. W. and North, A. M. (1959). *J. Amer. chem. Soc.* **81,** 1339.
Bevington, J. C. (1954). *J. chem. Soc.* 3707.
Bevington, J. C. (1955). *Trans. Faraday Soc.* **51,** 1392.
Bevington, J. C. (1956). *J. chem. Soc.* 1127.
Bevington, J. C. (1957). *Proc. roy. Soc.* A, **239,** 420.
Bevington, J. C. (1958). *J. Polym. Sci.* **29,** 235.
Bevington, J. C. (1960). Not yet published.
Bevington, J. C. and Brooks, C. S. (1956). *J. Polym. Sci.* **22,** 257.
Bevington, J. C. and Brooks, C. S. (1958). *Makromol. Chem.* **28,** 173.
Bevington, J. C. and Eaves, D. E. (1959). *Trans. Faraday Soc.* **55,** 1777.
Bevington, J. C. and Ghanem, N. A. (1956). *J. chem. Soc.* 3506.
Bevington, J. C. and Ghanem, N. A. (1958). *J. chem. Soc.* 2254.
Bevington, J. C. and Ghanem, N. A. (1959). *J. chem. Soc.* 2071.
Bevington, J. C., Ghanem, N. A. and Melville, H. W. (1955*a*). *Trans. Faraday Soc.* **51,** 346.
Bevington, J. C., Ghanem, N. A. and Melville, H. W. (1955*b*). *J. chem. Soc.* 2822.
Bevington, J. C., Guzman, G. M. and Melville, H. W. (1954). *Proc. roy. Soc.* A, **221,** 437, 453.
Bevington, J. C. and Lewis, T. D. (1958). *Trans. Faraday Soc.* **54,** 1340.
Bevington, J. C. and Lewis, T. D. (1960*a*), *Polymer*, **1,** 1.
Bevington, J. C. and Lewis, T. D. (1960*b*). Not yet published.
Bevington, J. C., Melville, H. W. and Taylor, R. P. (1954*a*). *J. Polym. Sci.* **12,** 449.
Bevington, J. C., Melville, H. W. and Taylor, R. P. (1954*b*). *J. Polym. Sci.* **14,** 463.
Bevington, J. C. and Toole, J. (1958). *J. Polym. Sci.* **28,** 413.
Bevington, J. C., Toole, J. and Trossarelli, L. (1958*a*). *Trans. Faraday Soc.* **54,** 863.
Bevington, J. C., Toole, J. and Trossarelli, L. (1958*b*). *Makromol. Chem.* **28,** 237.
Bevington, J. C., Toole, J. and Trossarelli, L. (1959). *Makromol. Chem.* **32,** 57.
Bevington, J. C. and Troth, H. G. (1960). Not yet published.
Bickel, A. F. and Waters, W. A. (1950*a*). *Rec. Trav. chim. Pays-Bas,* **69,** 1490.
Bickel, A. F. and Waters, W. A. (1950*b*). *J. chem. Soc.* 1764.
Blackley, D. C., Melville, H. W. and Valentine, L. (1954). *Proc. roy. Soc.* A, **227,** 10.

Blauer, G. (1953). *J. Polym. Sci.* **11**, 189.
Blauer, G. (1960). *Trans. Faraday Soc.* **56**, 606.
Blomquist, A. T. and Buselli, A. J. (1951). *J. Amer. chem. Soc.* **73**, 3883.
Bolland, J. L. and ten Have, P. (1947). *Disc. Faraday. Soc.* **2**, 252.
Bonsall, E. P., Valentine, L. and Melville, H. W. (1951). *J. Polym. Sci.* **7**, 39.
Bonsall, E. P., Valentine, L. and Melville, H. W. (1953). *Trans. Faraday Soc.* **49**, 686.
Boozer, C. E., Hammond, G. S., Hamilton, C. E. and Sen, J. N. (1955). *J. Amer. chem. Soc.* **77**, 3233, 3238.
Bosworth, P., Masson, C. R., Melville, H. W. and Peaker, F. W. (1952). *J. Polym. Sci.* **9**, 565.
Bovey, F. A. (1960). *J. Polym. Sci.* **46**, 59.
Bovey, F. A., Kolthoff, I. M., Medalia, A. I. and Meehan, E. J. (1955). "Emulsion Polymerizations". Interscience Publishers Inc., New York.
Bradbury, J. H. and Melville, H. W. (1954). *Proc. roy. Soc.* A, **222**, 456.
Braude, E. A., Brook, A. G. and Linstead, R. P. (1954). *J. chem. Soc.* 3574.
Breitenbach, J. W., Olaj, O. F. and Schindler, A. (1957). *Mh. Chem.* **88**, 1115.
Breitenbach, J. W. and Tschamler, H. (1951). *Mh. Chem.* **82**, 179.
Bristow, G. M. and Dainton, F. S. (1955). *Proc. roy. Soc.* A, **229**, 509, 525.
Brown, C. J. and Farthing, A. C. (1953). *J. chem. Soc.* 3270.
Burnett, G. M. (1954). "Mechanism of Polymer Reactions". Interscience Publishers Inc., New York.
Burnett, G. M., Evans, P. and Melville, H. W. (1953). *Trans. Faraday Soc.* **49**, 1096, 1105.
Burnett, G. M., George, M. H. and Melville, H. W. (1955). *J. Polym. Sci.* **16**, 31.
Burnett, G. M. and Gersmann, H. R. (1958). *J. Polym. Sci.* **28**, 655.
Burnett, G. M. and Loan, L. D. (1955). *Trans. Faraday Soc.* **51**, 214, 219, 226.
Bywater, S. (1955). *Trans. Faraday Soc.* **51**, 1267.
Carlin, R. B. and Shakespeare, N. E. (1946). *J. Amer. chem. Soc.* **68**, 876.
Chapiro, A., Magat, M., Sebban, J. and Wahl, P. (1955). *Ric. sci.* **25A**, 73.
Chen, C. S. H., Colthup, N., Deichert, W. and Webb, R. L. (1960). *J. Polym. Sci.* **45**, 247.
Chinmayanandam, B. R. and Melville, H. W. (1954). *Trans. Faraday Soc.* **50**, 73.
Cohen, S. G. and Sparrow, D. B. (1950). *J. Amer. chem. Soc.* **72**, 611.
Cohen, S. G. and Wang, C. H. (1955). *J. Amer. chem. Soc.* **77**, 2457.
Coleman, B. D. (1958). *J. Polym. Sci.* **31**, 155.
Condon, F. E. (1953). *J. Polym. Sci.* **11**, 139.
Cook, R. E. and Ivin, K. J. (1957). *Trans. Faraday Soc.* **53**, 1132.
Cooper, W. (1952). *J. Chem. Soc.* 2408.
Cotman, J. D. (1953). *Ann. N.Y. Acad. Sci.* **57**, 417.
Dainton, F. S., Diaper, J., Ivin, K. J. and Sheard, D. R. (1957). *Trans. Faraday Soc.* **53**, 1269.
Dainton, F. S. and Ivin, K. J. (1952). *Proc. roy. Soc.* A, **212**, 207.
Dainton, F. S. and Ivin, K. J. (1958). *Quart. Rev. chem. Soc., Lond.* **12**, 61.
Dainton, F. S., Ivin, K. J. and Walmsley, D. A. G. (1960). *Trans. Faraday Soc.* **56**, 1784.
Dainton, F. S. and James, D. G. L. (1958). *Trans. Faraday Soc.* **54**, 649.
Dainton, F. S. and Tordoff, M. (1957). *Trans. Faraday Soc.* **53**, 666.
Davis, P., Evans, M. G. and Higginson, W. C. E. (1951). *J. chem. Soc.* 2563.
Doak, K. W. (1948). *J. Amer. chem. Soc.* **70**, 1525.

Dogadkin, B. A. and Shernshnev, V. A. (1959) *Vysokomol. Soedin Vsesoyuz. Khim. Obshchestvo im. D. I. Mendeleeva*, **1**, 58.
Entwistle, E. R. (1960). *Trans. Faraday Soc.* **56**, 284.
Eriksson, A. F. V. (1956). *Svensk kem. Tidskr.* **68**, 301.
Errede, L. A. and Hopwood, S. L. (1957). *J. Amer. chem. Soc.* **79**, 6507.
Errede, L. A. and Szwarc, M. (1958) *Quart. Rev. chem. Soc., Lond.* **12**, 301.
Evans, M. G., Gergely, J. and Seaman, E. C. (1948). *J. Polym. Sci.* **3**, 866.
Ewald, A. H. (1956). *Disc. Faraday Soc.* **22**, 138.
Farmer, E. H. and Moore, C. G. (1951). *J. chem. Soc.* 142.
Ferington, T. and Tobolsky, A. V. (1958). *J. Amer. chem. Soc.* **80**, 3215.
Ferington, T. E. and Tobolsky, A. V. (1955). *J. Amer. chem. Soc.* **77**, 4510.
Flory, P. J. (1953). "Principles of Polymer Chemistry". Cornell University Press.
Flory, P. J. and Leutner, F. S. (1948). *J. Polym. Sci.* **3**, 880.
Fordyce, R. G., Chapin, E. C. and Ham, G. E. (1948). *J. Amer. chem. Soc.* **70**, 2489.
Foster, F. C. and Binder, J. L. (1953). *J. Amer. chem. Soc.* **75**, 2910.
Fox, T. G., Goode, W. E., Gratch, S., Huggett, C. M., Kincaid, J. F., Spell, A. and Stroupe, J. D. (1958). *J. Polym. Sci.* **31**, 173.
Fraenkel, G. K., Hirshon, J. M. and Walling, C. (1954). *J. Amer. chem. Soc.* **76**, 3606.
Frey, H. M. (1959). *Proc. chem. Soc., Lond.* 385.
Fuhrman, N. and Mesrobian, R. B. (1954). *J. Amer. chem. Soc.* **76**, 3281.
Funt, B. L. and Williams, F. D. (1960). *J. Polym. Sci.* **46**, 139.
Gee, G. and Melville, H. W. (1944). *Trans. Faraday Soc.* **40**, 240.
George, M. H., Grisenthwaite, R. J. and Hunter, R. F. (1958). *Chem. & Ind.* 1114.
Glavis, F. J. (1959). *J. Polym. Sci.* **36**, 547.
Gleason, E. H., Miller, M. L. and Sheats, G. F. (1959). *J. Polym. Sci.* **38**, 133.
Gluckman, M. S., Kampf, M. J., O'Brien, J. L., Fox, T. G. and Graham, R. K. (1959). *J. Polym. Sci.* **37**, 411.
Grassie, N. and Vance, E. (1956). *Trans. Faraday. Soc.* **52**, 727.
Gray, P. and Williams, A. (1959). *Chem. Rev.* **59**, 239.
Greene, F. D. (1956). *J. Amer. chem. Soc.* **78**, 2246.
Gregg, R. A. and Mayo, F. R. (1947). *Disc. Faraday Soc.* **2**, 328.
Grisenthwaite, R. J. and Hunter, R. F. (1958). *Chem. & Ind.* 719.
Haines, R. M. and Waters, W. A. (1958). *J. chem. Soc.* 3221.
Ham, G. E. (1954). *J. Polym. Sci.* **14**, 87.
Ham, G. E. (1960). *J. Polym. Sci.* **46**, 169, 177, 186.
Ham, G. E. and Ringwald, E. L. (1952). *J. Polym. Sci.* **8**, 91.
Hammond, G. S., Sen, J. N., and Boozer, C. E. (1955). *J. Amer. chem. Soc.* **77**, 3244.
Hammond, G. S., Trapp, O. B., Keys, R. T. and Neff, D. L. (1959). *J. Amer. chem. Soc.* **81**, 4878.
Harkins, W. D. (1947). *J. Amer. chem. Soc.* **69**, 1428.
Haward, R. N. (1950). *Trans. Faraday Soc.* **46**, 204.
Haward, R. N. and Simpson, W. (1951). *Trans. Faraday Soc.* **47**, 212.
Hawkins, E.G.E. (1950). *Quart. Rev. chem. Soc., Lond.* **4**, 251.
Hayden, P. and Melville, Sir H. W. (1960). *J. Polym. Sci.* **43**, 201, 215.
Henglein, A. (1955). *Makromol. Chem.* **15**, 188.
Henrici-Olivé, G. and Olivé, S. (1960). *Makromol. Chem.* **37**, 71.
Hey, D. H. and Misra, G. S. (1947). *Disc. Faraday Soc.* **2**, 279.
Hicks, J. A. (1956). *Trans. Faraday Soc.* **52**, 1526.
Huisgen, R. and Horeld, G. (1949). *Liebigs Ann.* **562**, 137.
Inamoto, N. and Simamura, O. (1958). *J. org. Chem.* **23**, 408.

Ingram, D. J. E. (1958). "Free Radicals as Studied by Electron Spin Resonance". Butterworth & Co. (Publishers) Ltd., London.
Ingram, D. J. E., Symons, M. C. R. and Townsend, M. G. (1958). *Trans. Faraday Soc.* **54,** 409.
Ivin, K. J. (1955). *Trans. Faraday Soc.* **51,** 1273.
Ivin, K. J., Keith, W. A. and Mackle, H. (1959). *Trans. Faraday Soc.* **55,** 262.
Jaffé, H. H. (1953). *Chem. Rev.* **53,** 191.
Jarkovsky, I., Stannett, V. and Szwarc, M. (1955). *J. Polym. Sci.* **18,** 515.
Jenkins, A. D. (1958). *Trans. Faraday Soc.* **54,** 1885, 1895.
Jenkins, A. D. and Johnston, R. (1959). *J. Polym. Sci.* **39,** 81.
Katagiri, K. and Okamura, S. (1955). *J. Polym. Sci.* **17,** 309.
Katagiri, K., Uno, K. and Okamura, S. (1955). *J. Polym. Sci.* **17,** 142.
Katchalsky, A. and Blauer, G. (1951). *Trans. Faraday Soc.* **47,** 1360.
Kharasch, M. S., Arimoto, F. S. and Nudenberg, W. (1951). *J. org. Chem.* **16,** 1556.
Kharasch, M. S., Fono, A. and Nudenberg, W. (1950). *J. org. Chem.* **15,** 763.
Kice, J. L. (1954). *J. Amer. chem. Soc.* **76,** 6274.
Kice, J. L. (1956). *J. Polym. Sci.* **19,** 123.
Kolthoff, I. M. and Miller, I. K. (1951). *J. Amer. chem. Soc.* **73,** 3055.
Kolthoff, I. M., O'Connor, P. R. and Hansen, J. L. (1955). *J. Polym. Sci.* **15,** 459.
Laird, R. K. (1956). *Disc. Faraday Soc.* **22,** 147.
Laita, Z. (1959). *J. Polym. Sci.* **38,** 247.
Laita, Z. and Macháček, Z. (1959). *J. Polym. Sci.* **38,** 459.
Landler, Y. (1952). *J. Polym. Sci.* **8,** 63.
Leffler, J. E. (1950). *J. Amer. chem. Soc.* **72,** 67.
Leffler, J. E. and Bond, W. B. (1956). *J. Amer. chem. Soc.* **78,** 335.
Lewis, F. M. and Mayo, F. R. (1948). *J. Amer. chem. Soc.* **70,** 1533.
Lím, D. and Wichterle, O. (1958). *J. Polym. Sci.* **29,** 579.
Litt, M. and Stannett, V. (1960). *Makromol. Chem.* **37,** 19.
Lowry, G. G. (1958). *J. Polym. Sci.* **31,** 187.
Lyons, J. A. and Watson, W. F. (1955). *J. Polym. Sci.* **18,** 141.
McBay, H. C. and Tucker, O. (1954). *J. org. Chem.* **19,** 869.
McBay, H. C., Tucker, O. and Milligan, A. (1954). *J. org. Chem.* **19,** 1003.
MacKie, J. S. and Bywater, S. (1957). *Canad. J. Chem.* **35,** 570.
Martin, J. T. and Norrish, R. G. W. (1953). *Proc. roy. Soc.* A, **220,** 322.
Marvel, C. S. and Wilson, B. D. (1958). *J. org. Chem.* **23,** 1479.
Marvel, C. S. and Woolford, R. G. (1958). *J. Amer. chem. Soc.* **80,** 830.
Mayo, F. R. (1953). *J. Amer. chem. Soc.* **75,** 6133.
Mayo, F. R., Miller, A. A. and Russell, G. A. (1958). *J. Amer. chem. Soc.* **80,** 2500.
Mayo, F. R. and Walling, C. (1950), *Chem. Rev.* **46,** 191.
Melville, H. W. (1956). *Proc. roy. Soc.* A, **237,** 149.
Melville, H. W., Peaker, F. W. and Vale, R. L. (1958*a*). *J. Polym. Sci.* **30,** 29.
Melville, H. W., Peaker, F. W. and Vale, R. L. (1958*b*). *Makromol. Chem.* **28,** 140.
Melville, H. W. and Sewell, P. R. (1959). *Makromol. Chem.* **32,** 139.
Melville, H. W. and Valentine, L. (1950). *Proc. roy. Soc.* A, **200,** 358.
Merrett, F. M. and Norrish, R. G. W. (1951). *Proc. roy. Soc.* A, **206,** 309.
Merz, E., Alfrey, T. and Goldfinger, G. (1946). *J. Polym. Sci.* **1,** 75.
Mochel, W. E., Crandall, J. L. and Peterson, J. H. (1955). *J. Amer. chem. Soc.* **77,** 494.
Molyneux, P. (1960). *Makromol. Chem.* **37,** 165.
Morrell, A. G. (1956). *Disc. Faraday Soc.* **22,** 152.
Morton, M., Salatiello, P. P. and Landfield, H. (1952). *J. Polym. Sci.* **8,** 215.

Nandi, U. S. and Palit, S. R. (1955). *J. Polym. Sci.* **17**, 65.
Nandi, U. S. and Palit, S. R. (1960). *Nature, Lond.* **185**, 235.
Nicholson, A. E. and Norrish, R. G. W. (1956). *Disc. Faraday Soc.* **22**, 97, 104.
Nicholson, A. J. C. (1954). *Trans. Faraday Soc.* **50**, 1067.
Norrish, R. G. W. and Simons, J. P. (1959). *Proc. roy. Soc.* A, **251**, 4.
Norrish, R. G. W. and Smith, R. R. (1942). *Nature, Lond.* **150**, 336.
Noyes, R. M. (1955). *J. Amer. chem. Soc.* **77**, 2042.
Nozaki, K. and Bartlett, P. D. (1946). *J. Amer. chem. Soc.* **68**, 1686.
O'Brien, E. L., Beringer, F. M. and Mesrobian, R. B. (1957). *J. Amer. chem. Soc.* **79**, 6238.
O'Driscoll, K. F. and Tobolsky, A. V. (1958). *J. Polym. Sci.* **31**, 123.
Offenbach, J. A. and Tobolsky, A. V. (1957). *J. Amer. chem. Soc.* **79**, 278.
Onyon, P. F. (1956). *Trans. Faraday Soc.* **52**, 80.
Orr, R. J. and Williams, H. L. (1955). *J. Amer. chem. Soc.* **77**, 3715.
Oster, G. (1954). *Nature, Lond.* **173**, 300.
Otsu, T. (1957). *J. Polym. Sci.* **26**, 236.
Otsu, T. and Nayatani, K. (1958). *Makromol. Chem.* **27**, 149.
Otsu, T., Nayatani, K., Muto, I. and Imai, M. (1958). *Makromol. Chem.* **27**, 142.
Overberger, C. G., Biletch, H., Finestone, A. B., Lilker, J. and Herbert, J. (1953). *J. Amer. chem. Soc.* **75**, 2078.
Overberger, C. G. and Lapkin, M. (1955). *J. Amer. chem. Soc.* **77**, 4651.
Palit, S. R. (1955). *Trans. Faraday Soc.* **51**, 1129.
Palit, S. R. (1960). *Makromol. Chem.* **38**, 96.
Palit, S. R. and Das, S. K. (1954). *Proc. roy. Soc.* A, **226**, 82.
Pinner, S. H. (1952). *J. Polym. Sci.* **9**, 282.
Poirier, R. H., Kahler, E. J. and Benington, F. (1952). *J. org. Chem.* **17**, 1437.
Price, C. C. (1948). *J. Polym. Sci.* **3**, 772.
Rembaum, A. and Szwarc, M. (1954). *J. Amer. chem. Soc.* **76**, 5975.
Rembaum, A. and Szwarc, M. (1955*a*). *J. Chem. Phys.* **23**, 909.
Rembaum, A. and Szwarc, M. (1955*b*). *J. Amer. chem. Soc.* **77**, 3486.
Reynolds, W. B. and Cotten, E. W. (1950). *Industr. Engng Chem.* **42**, 1905.
Richardson, W. S. and Sacher, A. (1953). *J. Polym. Sci.* **10**, 353.
Robb, J. C. and Vofsi, D. (1959). *Trans. Faraday Soc.* **55**, 558.
Robertson, E. R. (1956). *Trans. Faraday Soc.* **52**, 426.
Roedel, M. J. (1953). *J. Amer. chem. Soc.* **75**, 6110.
Russell, G. A. (1956). *J. Amer. chem. Soc.* **78**, 1047.
Russell, G. A. (1959). *J. org. Chem.* **24**, 300.
Russell, K. E. (1955). *J. Amer. chem. Soc.* **77**, 4814.
Russell, K. E. and Tobolsky, A. V. (1953). *J. Amer. chem. Soc.* **75**, 5052.
Russell, K. E. and Tobolsky, A. V. (1954). *J. Amer. chem. Soc.* **76**, 395.
Schröder, G. (1958). *J. Polym. Sci.* **31**, 309.
Schulz, G. V. and Blaschke, F. (1942). *Z. phys. Chem.* 1942, **51**, B, 75.
Schulz, G. V. and Henrici, G. (1956). *Makromol. Chem.* **18**, 437.
Schulz, G. V. Henrici, G. and Olivé, S. (1955). *J. Polym. Sci.* **17**, 45.
Schulz, G. V., Henrici, G. and Olivé, S. (1956). *Z. Elektrochem.* **60**, 296.
Schulz, G. V., Henrici-Olivé, G. and Olivé, S. (1959). *Makromol. Chem.* **31**, 88.
Sengupta, R. and Palit, S. R. (1951). *J. chem. Soc.* 3278.
Shah, H. A., Leonard, F. and Tobolsky, A. V. (1951). *J. Polym. Sci.* **7**, 537.
Shelton, J. R., McDonel, E. T. and Crano, J. C. (1960). *J. Polym. Sci.* **42**, 289.
Small, P. A. (1953). *Trans. Faraday Soc.* **49**, 441.
Smets, G. (1957). *Coll. Trav. Chim. Tchécosl.* **22**, 264.

Smets, G. and Hayashi, K. (1958). *J. Polym. Sci.* **29**, 257.

Smets, G., Poot, A., Mullier, M. and Bex, J. P. (1959). *J. Polym. Sci.* **34**, 287.

Smid, J. and Szwarc, M. (1956). *J. Amer. chem. Soc.* **78**, 3322.

Smith, P. and Carbone, S. (1959). *J. Amer. chem. Soc.* **81**, 6174.

Smith, P. and Rosenberg, A. M. (1959). *J. Amer. chem. Soc.* **81**, 2037.

Smith, W. V. and Ewart, R. H. (1948). *J. chem. Phys.* **16**, 592.

Stockmayer, W. H. (1945). *J. chem. Phys.* **13**, 199.

Stockmayer, W. H., Howard, R. O. and Clarke, J. T. (1953). *J. Amer. chem. Soc.* **75**, 1756.

Stockmayer, W. H. and Peebles, L. H. (1953). *J. Amer. chem. Soc.* **75**, 2278.

Strause, S. F. and Dyer, E. (1956). *J. Amer. chem. Soc.* **78**, 136.

Suzuki, M., Miyama, H. and Fujimoto, S. (1959). *J. Polym. Sci.* **37**, 533.

Swain, C. G., Stockmayer, W. H. and Clarke, J. T. (1950). *J. Amer. chem. Soc.* **72**, 5426.

Szwarc, M. (1955). *J. Polym. Sci.* **16**, 367.

Szwarc, M., Levy, M. and Milkovich, R. (1956). *J. Amer. chem. Soc.* **78**, 2656.

Talât-Erben, M. and Bywater. S. (1955*a*). *Ric. Sci.* **25A**, 11.

Talât-Erben, M. and Bywater, S. (1955*b*). *J. Amer. chem. Soc.* **77**, 3710.

Talât-Erben, M. and Isfendiyaroğlu, A. N. (1958). *Canad. J. Chem.* **36**, 1156.

Talât-Erben, M. and Isfendiyaroğlu, A. N. (1959). *Canad. J. Chem.* **37**, 1165.

Talât-Erben, M. and Önol, N. (1960). *Canad J. Chem.* **38**, 1154.

Thomas, W. M., Gleason, E. H. and Pellon, J. J. (1955). *J. Polym. Sci.* **17**, 275.

Tobolsky, A. V. and Baysal, B. (1953). *J. Amer. chem. Soc.* **75**, 1757.

Tobolsky, A. V. and Mesrobian, R. B. (1954). "Organic Peroxides". Interscience Publishers Inc., New York.

Trotman-Dickenson, A. F. (1958). *Ann. Rep. Progr. Chem.* **55**, 36.

Trotman-Dickenson, A. F. (1959). "Free Radicals". Methuen & Co. Ltd., London.

Tüdös, F. (1958). *J. Polym. Sci.* **30**, 343.

Uri, N. (1952). *Chem. Rev.* **50**, 375.

Van Hook, J. P. and Tobolsky, A. V. (1958). *J. Amer. chem. Soc.* **80**, 779.

Vaughan, M. F. (1952). *Trans. Faraday Soc.* **48**, 576.

Verdin, D. (1960). *Trans. Faraday Soc.* **56**, 823.

Wall, L. A. (1947). *J. Polym. Sci.* **2**, 542.

Wall, L. A. and Brown, D. W. (1954). *J. Polym. Sci.* **14**, 513.

Walling, C. (1945). *J. Amer. chem. Soc.* **67**, 441.

Walling, C. (1948). *J. Amer. chem. Soc.* **70**, 2561.

Walling, C. (1954). *J. Polym. Sci.* **14**, 214.

Walling, C. (1955). *J. Polym. Sci.* **16**, 315.

Walling, C. (1957). "Free Radicals in Solution". John Wiley & Sons Inc., New York.

Walling, C. and Briggs, E. R. (1945). *J. Amer. chem. Soc.* **67**, 1774.

Walling, C., Briggs, E. R. and Mayo, F. R. (1946). *J. Amer. chem. Soc.* **68**, 1145.

Walling, C. and Chang, Y. W. (1954). *J. Amer. chem. Soc.* **76**, 4878.

Walling, C. and Savas, E. S. (1960). *J. Amer. chem. Soc.* **82**, 1738.

Walling, C., Seymour, D. and Wolfstirn, K. B. (1948). *J. Amer. chem. Soc.* **70**, 1544.

Whiffen, D. H. (1959). *Makromol. Chem.* **34**, 170.

Worsfold, D. J. and Bywater, S. (1957). *J. Polym. Sci.* **26**, 299.

Zand, R. and Mesrobian, R. B. (1955). *J. Amer. chem. Soc.* **77**, 6523.

AUTHOR INDEX

SUBJECT INDEX